ON THE ORIGIN OF EVOLUTION

TRACING 'DARWIN'S DANGEROUS IDEA' FROM ARISTOTLE TO DNA

JOHN AND MARY GRIBBIN

WILLIAM
COLLINS

William Collins
An imprint of HarperCollins*Publishers*
1 London Bridge Street
London SE1 9GF

WilliamCollinsBooks.com

HarperCollins*Publishers*
1st Floor, Watermarque Building, Ringsend Road
Dublin 4, Ireland

First published in Great Britain by William Collins in 2020
This William Collins paperback published in 2021

2022 2024 2023 2021
10 9 8 7 6 5 4 3 2 1

A catalogue record for this book is available from the British Library

ISBN 978-0-00-833340-9

Typeset in Stempel Garamond by
Palimpsest Book Production Ltd, Falkirk, Stirlingshire
Printed and Bound in the UK using
100% Renewable Electricity at CPI Group (UK) Ltd

MIX
Paper from
responsible sources

FSC
www.fsc.org
FSC™ C007454

This book is produced from independently certified FSC™ paper
to ensure responsible forest management.

For more information visit: www.harpercollins.co.uk/green

EVOLUTION IN A NUTSHELL

The process of evolution by natural selection requires that living things reproduce, making copies of themselves, but that the copies are not quite perfect, so that there is variety in the next generation. If that variety makes some of those offspring more successful than others at reproducing in their turn, for whatever reason, the characteristics that make them more successful will spread among subsequent generations – they will be selected.

But in order to take part in the selection, you have to live long enough to reproduce, and if you live longer and reproduce more, all to the good. This has led to a neat one-line summary of Darwin's theory: dead animals have less chance of reproducing than live ones. Or, if you prefer: live long and prosper.

CONTENTS

PREFACE

EXPLODING THE DARWIN MYTH

The scene is a study in a country house in a Kent village. Over the past twenty years, the man sitting at the desk has been quietly gathering evidence to support his revolutionary theory about the origins of life. Only a few close friends know about his work, and he is not yet ready to announce it to the world at large. The post is brought in by a servant. One particular letter catches his eye; although securely fastened, it is travel-stained and has clearly come a long way. He picks up a paper knife, opens the package, and begins to read. The knife, forgotten, drops on the desk; he feels dizzy, heart pounding. He has been pre-empted. Some unknown plant hunter on the other side of the world has somehow stumbled upon the same great idea. There is nothing for it but to abandon any claims to priority and accept his fate as a footnote to history. If only he had published!

That, at any rate, is more or less the popular myth of how Charles Darwin discovered that Alfred Russel Wallace had independently developed the theory of evolution by natural selection. And it is more or less wrong. In that popular myth, Charles Darwin was a lone genius who spent decades puzzling over the origin of species in his secluded country retreat, before a Eureka! moment when everything fell into place and he saw what nobody else had been bold enough to see – that species evolve. In reality, by 1859, when Darwin's great book on this theory was published, the *fact* of evolution was widely accepted, and had been discussed seriously by scientists for decades. Darwin's special contribution, though, was to explain the mechanism of evolution, the process of natural selection that enables individuals that are better suited to thrive and produce offspring, while those that are less 'fit' struggle and leave fewer offspring. But even this was not a *unique* insight. The same idea occurred independently to another naturalist, Alfred Russel Wallace, and Darwin was forced to go public with his idea sooner than intended when he received a letter from Wallace setting out the idea. That much is true. It is a sign of how insidiously the popular myth has spread that even a respectable historian of science working at the University of Cambridge could write, as recently as 2009, that:

> When a letter arrived from an unknown collector in Malaysia, Darwin realised that other people were thinking along similar lines. Protected by his allies, he warded off this potential rival and rushed into print with On the Origin of Species.[1]

Wallace was not an unknown collector, but a naturalist in his own right. He created collections and sold them to fund his travels, because unlike Darwin he was not lucky enough to have inherited wealth. He was a regular correspondent with

Darwin, and had published scientific papers on 'the species problem', one of which appeared in print in 1855 and prompted Darwin to mention in a letter to Wallace that 'I can plainly see that we have thought much alike'. When Darwin realised just how much alike, his instinct was not to ward off his rival, but to let Wallace publish and take all the credit. As we describe in Chapter Six, it was Darwin's friends who had the inspired idea of putting together a joint scientific paper under both their names, to present the idea of natural selection to the world at large and ensure that credit was shared.

By 1859, evolution by natural selection was an idea whose time had come, and if neither Darwin nor Wallace had come up with it somebody else very soon would have – perhaps Wallace's friend Henry Bates, who also features in our story. But how did that situation come about, and why is it that the origin story celebrates Darwin, not Wallace? That is what this book is all about.

INTRODUCTION

Evolution is a fact. It is observed to happen in nature (most famously among the finches of the Galapagos Islands studied by Charles Darwin), in the fossil record of life on Earth, and in the way 'superbugs' evolve resistance to antibiotics. Theories are put forward to explain this fact, in the same way that Isaac Newton and Albert Einstein put forward theories of gravity to explain the fact that things fall down and the planets are held in orbit around the Sun. The best theory of gravity we have is Einstein's general theory of relativity, in the sense that it does a good job of explaining the observed facts, although Newton's theory is pretty good for many purposes. The best theory of evolution we have, in the sense that it does a good job of explaining the observed facts, is the theory of natural selection, although possibly it may not be the last word on the subject, just as Newton's theory of gravity was not the last word on that subject. But apples did not stop falling off trees when Einstein improved on Newton's theory, and living things will not stop evolving if someone improves on Darwin's theory (which, spoiler alert, has indeed happened). So this book is not really about the origin of evolution, but about the origin of ideas about evolution, which would not have made such a snappy title.

As this implies, the theory of evolution by natural selection did not spring fully formed and unprecedented from the brain of Charles Darwin. The idea of evolution had been around, in various guises, since the time of Ancient Greece, and even natural selection, the key to Darwinian evolution, had been half-seen, as if through a glass darkly, by some of his predecessors and contemporaries, while one of those contemporaries, Alfred Russel Wallace, saw it as clearly as Darwin did. Nor did theorising about evolution stop with what Daniel Dennett called 'Darwin's Dangerous Idea'. Our aim is to put that idea in its proper context, showing how it built on what went before and how it was further developed in the twentieth century, through an understanding of genetics and the biochemical bases of evolution, into the so-called 'modern synthesis', and beyond. None of this diminishes the achievement of Darwin himself in perceiving the way evolution works at the level of individuals and species. The idea is obvious once it has been explained. As Thomas Henry Huxley, 'Darwin's bulldog', commented when he first learned of the theory, 'how extremely stupid not to have thought of that'. But, as is well known, hindsight has 20:20 vision, and it required great insight to be one of the first to think of that. Darwin's other great contribution, unlike, for example, Wallace, was to present the idea in a clear and accessible book that lesser mortals could understand. He deserves his recognition as the primary proponent of the idea of natural selection, but, as we hope to show, his contribution was one link in a chain that extends back into antiquity, and is still being forged today. Our story is incomplete, in the sense that we have not tried to describe the work of everyone who is on record as having thought about evolution, but by highlighting the main players we hope to provide an overview of how the story developed, before and after Darwin.

John Gribbin
Mary Gribbin
April 2019

PART ONE

ANCIENT TIMES

CHAPTER ONE

THROUGH A
GLASS, DARKLY

In nineteenth-century Europe, the idea of evolution was revolutionary because it overturned the established Christian tradition of an essentially unchanging world in which everything, including the forms of living things, had been fixed by God. This tradition actually predated Christianity. In Ancient Greece, Plato, his student Aristotle and the Stoics all taught that the forms of all living things were fixed by the gods. Plato's philosophy was based on the idea of 'essence'. He argued that the essence is the perfect embodiment of an object. There is, for example, an essential, perfect triangle, but any triangle we can draw on Earth is only an imperfect approximation to the essence. In the same way, each kind of plant or animal has a God-given essence. There is an essential horse, a perfect example of its kind, which incorporates all the characteristics of horsiness, but any living horse on Earth is only an imperfect representation of the essential horse, which is why horses differ from one another. But a horse could never be changed into, say, a zebra, any more than a triangle can be changed into a square.

Aristotle, who lived from 384 BCE to 322 BCE, developed this idea further, and was particularly influential on later generations of Christian thinkers because much of his writing was preserved. It was Aristotle who gave those thinkers the idea of a 'great chain of being', or a 'ladder of life', in which different kinds of life on Earth are placed in order of their complexity (with human beings, of course, at the top). He stated that the properties of living things showed that they had what he called a 'final cause', meaning that each variety had been designed for a purpose. But what is equally interesting for us is that Aristotle took the trouble to reject the ideas of his predecessor Empedocles (*c.* 490 BCE to 430 BCE), who had put forward the idea that the forms of living things might have originated by chance. If Empedocles was important enough to be noticed in this way, his ideas must have had some contemporary influence. This was not a theory of evolution, but in a sense it did involve the idea of natural selection. Aristotle explained the process, but only in order to say that it was nonsense and absurd. After pointing out how front teeth are sharp and adapted to cutting food, while back teeth are broad and adapted to grinding, he wrote that it 'may be said' that this is not by design, but because the:

> . . . *arrangement came about by chance; and the same reasoning is applied to other parts of the body in which existence for some purpose is apparent. And it is argued that where all things happened as if they were made for some purpose, being aptly united by chance, these were preserved, but such as were not aptly made, these were lost and still perish, according to what Empedocles says.*[2]

Aristotle dismisses this as 'impossible' – such things could not be produced 'by fortune or chance'. But what he is actually dismissing here is the idea that front and back teeth should suddenly appear, among a variety of different kinds of teeth, with only the most 'apt' surviving. What the ancients failed to grasp is the gradualness of evolution, the way small changes build up over many generations. A closer look at what Empedocles said highlights this.

Empedocles' ideas have only come down to us in fragments of his writing, and in references to his work made by other writers. The fragments have been collected and translated by William Leonard (published in 1908), and they give us a glimpse of Empedocles' vision of a primordial origin of life in which grotesque combinations of heads, bodies, eyes and limbs were joined at random:

> *There budded many a head without a neck,*
> *And arms were roaming, shoulderless and bare,*
> *And eyes that wanted foreheads drifted by*
> . . .
> *In isolation wandered every limb,*
> *Hither and thither seeing union meet*
> . . .
> *These members fell together where they met,*
> *And many a birth besides was then begot*
> *In a long line of ever varied life.*
> . . .
> *Creatures of countless hands and trailing feet.*
> . . .
> *Many were born with twofold brow and breast,*
> *Some with the face of man on bovine stock,*
> *Some with man's form beneath a bovine*
>   *head,*
> *Mixed shapes of being . . .*

But only the forms best suited for life survived and reproduced. Although according to his scenario all this happened long ago, there is a hint in his writings that Empedocles believed that some form of evolution might continue in the present day, because living creatures are still imperfect.

An earlier Greek philosopher, Anaximander (c. 610 BCE to 546 BCE), is regarded as one of the first proponents of the scientific approach to nature, trying to explain different aspects of the world by assuming that nature is ruled by laws. As with Empedocles, very little of his writing survives, but we learn from later authors of a particularly perceptive insight. Anaximander pointed out that because human beings have an extended infancy and are helpless when young, the first humans could not have appeared as unprotected babies. His solution to the puzzle was that fish emerged in the primordial ocean before humans, but that the first humans developed in some way inside fish-like creatures, in a kind of capsule floating in the water, where they could grow until puberty before, like a butterfly emerging from a chrysalis, they burst out as adults capable of looking after themselves. There is more than a hint here of the idea that the first human beings were not created fully formed.

Epicurus (341 BCE to 270 BCE) leaned more towards Empedocles' version of events, complete with monstrosities. He was a materialist who saw no role for gods. In his view, the first creatures formed through combinations of atoms, and those that were best at surviving did, while others did not. His philosophy was propounded and developed by the Roman author Lucretius (c. 99 BCE to c. 55 BCE), whose poem *De Rerum Natura (The Nature of Things)* provides the best summary of this line of thinking of his Greek predecessors.

Lucretius was an atomist, who believed that the world is only a temporary arrangement of these fundamental particles (as we would now describe them). This was one of his arguments against there being a benevolent creator, because such a being would, he argued, have ensured that his creation would last forever (it's worth noting that Plato used the argument the other way round, saying that the world had been created by a benevolent god, so it must be everlasting). And, Lucretius pointed out, if the world has been made by a benevolent creator and designed for our benefit, why was it so hostile to human life? He also addressed the question of how life on Earth got started. The young Earth, he thought, was so fertile that life forms emerged spontaneously from the ground, in all kinds of random structures. Most of these died because they were unable to feed or reproduce, but a few kinds survived because they had strength or cunning, or (a sign that even Lucretius thought people were special) because they were useful to humankind. But he also emphasises that the creatures that survived had to be capable of reproducing their kind. There are clear elements here of the modern idea of evolution by natural selection. There must be a variety on which selection can act, and species must be able to reproduce successfully. But there is no suggestion that the process of reproduction might produce the variety on which selection could act. And the selection process is, once again, seen as something that happened long ago and has now stopped. The ancients did not have a theory of evolution as we now understand it, but some of them at least had the basis of an idea as to why different forms of life seem to be designed for their roles among the multitude of forms of life on Earth.

What we might think of as precursors of evolutionary ideas were also discussed in other cultures. In China, Zhuang Zhou (c. 369 BCE to 286 BCE), one of the founders of Taoist philosophy, referred to biological change. Taoism rejects the idea of fixed biological species, talking instead of 'constant transformation', and comes close to providing an image of the 'struggle for survival' – an image that independently influenced the thinking of both Darwin and Wallace. In the biological world, every species is the prey of another. Even creatures at the top of the food chain, such as lions, are 'preyed upon' by diseases. Taoists explain this lack of harmony by arguing that if there was a species that was not preyed upon in this way, it would reproduce unchecked, consuming all resources and leading to its own end. If human beings wipe out disease and continue to reproduce unchecked, they put themselves and the world in danger. A variation on this theme, expressed not by a Taoist philosopher but by the eighteenth-century English cleric Thomas Malthus, would influence the thinking of Darwin and Wallace.

Closer to home, both geographically and historically, Islamic scholars puzzled over the relationship between the living and non-living worlds, the interactions of different forms of life with one another, and the relationship between human beings and other animals. Aristotle was translated into Arabic in the first half of the ninth century, and in the tenth century what we would now call science became a matter of intense debate among the scholars in Spain, then part of the Islamic world. In the ninth century, al-Jahiz (776 CE to 868 CE) wrote in his *Book of Animals (Kitab al-Hayawan)*:

> *All animals, in short, cannot exist without food, neither can the hunting animal escape being hunted in his turn. Every weak animal devours those weaker than itself. Strong animals cannot escape being devoured by other animals stronger than they. And in this respect, men do not differ from animals, some with respect to others, although they do not arrive at the same extremes. In short, God has disposed some human beings as a cause of life for others, and likewise, he has disposed the latter as a cause of the death of the former.*[3]

Some of these scholars had the beginning of an understanding of the immense length of time for which the Earth, and life on Earth, had existed. The Persian polymath Avicenna (*c.* 980 CE to 1037 CE) wrote:

> *Mountains may be due to two causes. Either they are effects of upheavals of the crust of the Earth, such as might occur during a violent earthquake, or they are the effect of water, which, cutting for itself a new route, has denuded the valleys, the strata being of different kinds, some soft, some hard. The winds and waters disintegrate the one, but leave the other intact. Most of the eminences of the Earth have had this latter origin. It would require a long period of time for all such changes to be accomplished, during which the mountains themselves might be somewhat diminished in size.*[4]

In the thirteenth century, the Persian polymath Nasir al-Din al-Tusi (1201 to 1274) discussed the way in which organisms are adapted to their environments, using language that has sometimes been interpreted as describing a theory of evolution; but this seems to be to some extent wishful thinking. In his book *Akhlaq-i Nasiri*, Tusi dealt with a variety of biological topics and described his version of the ladder of life. His discussion of the origin of life echoes that of Lucretius, by starting in the chaos from which order and life originated, with some forms of life succeeding and others failing. The passage that excites people looking for early evolutionary thinking reads:

> *The organisms that can gain the new features faster are more variable. As a result, they gain advantages over other creatures. [. . .] The bodies are changing as a result of the internal and external interactions.*

But it is far from clear whether Tusi is suggesting that these changes are being acquired as one generation succeeds another, or whether an individual is changing its body in response to environmental stresses – the idea now known as Lamarckism, which we discuss later (Chapter Four).

This ambiguity also applies to the interpretation of the words of other Islamic scholars. In 1377, in his book *Al-Muqaddimah*, Ibn Khaldun (1332 to 1406) wrote:

The animal world then widens, its species become numerous, and, in a gradual process of creation, it finally leads to man, who is able to think and reflect. The higher stage of man is reached from the world of monkeys, in which both sagacity and perception are found, but which has not reached the stage of actual reflection and thinking. At this point we come to the first stage of man. This is as far as our [physical] observation extends.

It is not clear whether he was talking about the development of humans *from* monkeys, rather than simply placing species on the ladder of creation, but he does also refer to 'transformations of some existent things into others'.

This is enough to show that long before Darwin and well beyond the boundaries of Western Europe there were people who thought seriously about humankind's place in nature, and the relationship of living species to one another. The fact is, though, that our modern understanding of evolution did emerge from the Christian society of Western Europe, where its development was certainly not helped by the established religious environment. But things might have followed a different path even within the context of Christian thinking, if the Church followers had taken more notice of some of its early thinkers.

Some important figures in the early Christian Church

realised that the biblical account of the Creation in Genesis should not be taken literally, and saw that life on Earth must have developed in some way from more primitive origins, even if all this was guided by God. Origen of Alexandria (*c.* 184 to *c.* 253) was one of the most important early Christian philosopher-theologians, who produced an enormous body of work. This included his advocacy of the idea that the Bible story should be considered as an allegory, and not a literal account of the creation of the world. For his pains (not just for this idea), Origen was condemned as a heretic by a council at Alexandria in the year 400, and in 543, the Emperor Justinian I repeated the condemnation and ordered all his writings to be burned. As with Aristotle's denunciation of Empedocles, the fact that Justinian bothered to do this nearly three hundred years after Origen died shows how wide his influence was.

By the time Justinian was retrospectively censoring Origen, Bishop Augustine of Hippo (Saint Augustine, 354 CE to 430 CE) had made his own contribution to the debate about Genesis. Augustine was another prolific writer, and his ideas on various subjects changed over time, but one of his key teachings was that if a literal interpretation of the Bible conflicts with logic and reason (what he regarded as our God-given ability to reason, which made it all the more important), then the Bible story should be interpreted as metaphor or allegory. The stories had been written, he argued, in this simple form to make them intelligible to the people who lived at the time when Genesis was written. This suggestion comes in Book V of his epic work *De Genesi ad Litteram (On the Literal Interpretation of Genesis)*. He says that the correct interpretation of Genesis is that animals and plants emerge from water and earth and 'develop in time . . . each according to its nature.' He makes an analogy with the growth of a tree from a seed, emerging from the earth and developing into the mature form. But he does not make this analogy with the growth of, say, an animal from an embryo, but with species developing from simpler beginnings. God created the potentiality for living things, which were duly brought forth 'in the course of time on different days according to their different kinds'. This is change, but not really evolution, because everything is planned in advance by God. 'In accordance with those kinds of creatures which He first made, God makes many new things which He did not make then . . . God unfolds the generations which He laid up in creation when first He founded it.' Developing from his seed analogy, Augustine says:

> In the [seed] there is invisibly present all that will develop into the tree. And in this same way we must picture the [origin of the] world . . . This includes not only the Heaven with the Sun, Moon and stars . . . it includes also the beings which water and earth produced in potency and in their causes before they came forth in the course of time.

'Plant, fowl and animal life are not perfect,' says Augustine, 'but created in a state of potentiality.'

In another book, *De Genesi (On Genesis),* he writes, 'To suppose that God formed man from dust with bodily hands is very childish . . . God neither formed man with bodily hands nor did he breathe upon him with throat and lips.' Although his theology became one of the main pillars of the Church, somehow this aspect of Augustine's thinking was neglected in favour of the simplistic biblical account, promulgated for the benefit of the uneducated masses. What if things had been different? In his contribution to the debate in the second half of the nineteenth century following the publication of *On the Origin of Species,* Henry Osborn wrote, in his book *From the Greeks to Darwin*:

If the orthodoxy of Augustine had remained the teaching of the Church, the final establishment of Evolution would have come earlier than it did, certainly during the eighteenth instead of the nineteenth century, and the bitter controversy over this truth of Nature would never have arisen . . . Plainly as the direct or instantaneous Creation of animals and plants appeared to be taught in Genesis, Augustine reads this in the light of primary causation and the gradual development from the imperfect to the perfect of Aristotle. This most influential teacher thus handed down to his followers opinions which closely conform to the progressive views of those theologians of the present days who have accepted the Evolution theory.

Whether the views of those nineteenth-century theologians went far enough is another matter, if they simply saw evolution in terms of primary causation and the gradual development from the imperfect to the perfect of Aristotle.

Aristotle's ideas took root in the Western Christian Church after the twelfth century, when Latin translations of Islamic texts that were themselves translations of Ancient Greek texts became available to scholars. The most influential of these scholars was Thomas Aquinas (1225 to 1274), another saint. Although he did not agree with Augustine's interpretation of the seven days of creation as a metaphor, and believed it was literally true that God created the world in six ordinary days and rested on the seventh, he seems to have approved of much of what Augustine said, interpreting the story of Genesis as meaning that God stopped making new creatures on the seventh day in the sense that everything that came afterwards was not original, because it had ancestors in the same 'likeness' – what we might now think of as species. 'All things that were produced in the process of time through the work of divine providence, with creation operating under God, were produced in the first condition of things according to certain seminal patterns, as Augustine says . . . on the day on which God created Heaven and Earth, He also created every plant of the field, not, indeed, actually, but "before it sprung up in the earth", that is, potentially.' This allows for a kind of development of life on Earth as time passes, and even for individual species to improve in some sense as they strive for Aristotelian perfection. But, crucially, it explicitly rejects the idea of new species evolving since the creation.

Interestingly, although Thomas argued that God directly created every human soul, he seems to have had no difficulty reconciling this with the idea that humans are subject to the same rules of behaviour as other animals. Matt Rossano, of Southeastern Louisiana University, has pointed out the similarity between some of Thomas's teaching and the ideas of modern evolutionary psychology (what used to be known as sociobiology). In his *Summa Contra Gentiles*, Thomas writes:

> *We observe that in those animals, dogs for instance, in which the female by herself suffices for the rearing of offspring, the male and female stay no time together after the performance of the sexual act. But in all animals in which the female by herself does not suffice for the rearing of the offspring, male and female dwell together after the sexual act so long as is necessary for the rearing and training of the offspring. This appears in birds, whose young are incapable of finding their own food immediately after they are hatched . . . Hence, whereas it is necessary in all animals for the male to stand by the female for such time as the father's concurrence is requisite for bringing up the progeny, it is natural for man to be tied to the society of one fixed woman for a long period, not a short one.*

Thomas also understood the importance of what is now called 'paternity certainty' – reassuring a male that his own genes are being passed on to the next generation:

> *Every animal desires free enjoyment of pleasure of sexual union as of eating: which freedom is impeded by there being either several males to one female, or the other way about . . . But in men there is a special reason, inasmuch as man naturally desires to be sure of his own offspring . . . The reason why a wife is not allowed more than one husband at a time is because otherwise paternity would be uncertain.*

The question Thomas fails to ask is, *why* should paternity matter to the male, if everything is ordained by God? He sees this as a natural desire, but it hints – more than hints – at a profound feature of modern evolutionary theory, which addresses such questions as why 'natural' patterns of behaviour have evolved. It is all explained by modern evolutionary theory in terms of individual animals maximising their chances of passing copies of their genes on to the next generation – the copying that is such an important feature of evolution by natural selection. 'Natural' behaviour seems natural to us because it has been successful, in evolutionary terms. Without knowing anything about genes, Thomas clearly saw the reasons for such behaviour in the animal world, and equally clearly saw that there was no distinction, in this regard, between the behaviour of human beings and that of other animals. It is hard not to imagine that if someone as perceptive and intelligent as Thomas had been presented with the evidence that was available six centuries later to Charles Darwin, he would have accepted – or discovered for himself – the idea of evolution by natural selection, even if he held to his view of God as the creator of human souls. Unfortunately, for most of those six centuries most of the people in charge of the teaching of the Church were not as perceptive and intelligent as Thomas, and the official line was that the world we see around us is fixed and unchanging, designed by God. As far as life was concerned, the chain, or ladder, of being was the correct image. Each species had its place as a link in the chain, or a rung on the ladder, which actually extended all the way from God at the top down through the angels and humans (mostly mortal, but with a soul made of spirit) to animals, plants and minerals. This was a powerful image for rulers in the centuries following Thomas, because it could be extended to say that the place of every individual human being in society was ordained by God as part of the great chain of being. If you were a peasant or a noble, a beggar or a king, you simply had to accept your lot because that was the way God had ordered it. It would be sinful to debase yourself and behave like a lower animal – but just as sinful to get ideas above your station and think or act as if you were as good as someone of higher status. So the establishment had a vested interest in promoting the idea.

In this Christianised version of the Platonic/Aristotelian world, no species could ever move from one place in the chain to another, because there are no empty links and every link (every rung of the ladder) is occupied by one species, with species on adjacent steps closely resembling one another. This idea remained a central principle of biological thinking right into the eighteenth century. Nothing better sums up its influence than the words of Alexander Pope, published in 1714 in *The Rape of the Lock*:

> *Vast chain of being! which from God began,*
> *Natures ethereal, human, angel, man,*
> *Beast, bird, fish, insect, what no eye can see,*
> *No glass can reach; from Infinite to thee,*

From thee to nothing, – On superior pow'rs
Were we too press, inferior might on ours;
Or in the full creation leave a void,
Where, one step broken, the great scale's destroy'd;
From Nature's chain whatever link you strike,
Tenth, or ten thousandth, breaks the chain alike.

But by then the idea of biological evolution and the transformation of species had been clearly put forward by one of
the greatest geniuses of the seventeenth century, a key player
in the scientific revolution that had begun in the middle of
the previous century.

CHAPTER TWO

A FALSE DAWN

The revival of Western Europe known as the Renaissance has been linked with the collapse of the Eastern Roman Empire (Byzantium) in the fifteenth century and the movement of Greek-speaking scholars to Italy and further west, who took with them ideas and documents that encouraged a rebirth of civilisation. There were other factors, too, not least the development of moveable type by Johann Gutenberg in the same century, but whatever the reasons the Renaissance was well under way by the beginning of the sixteenth century.

At first, this intellectual flowering accepted the teaching of the ancients as the best description of the material and living worlds; Greeks such as Aristotle were regarded as intellectually superior to their sixteenth-century heirs, who were merely rediscovering things that the ancients already knew. But things soon began to change. A convenient date for the beginning of the scientific Renaissance is 1543, the year in which Nicolaus Copernicus published his book *De Revolutionibus orbium coelestium (On the Revolutions of the Heavenly Spheres)*, which said that the Earth goes round the Sun, and Andreas Vesalius published his equally important (but less famous) book *De humani corporis fabrica (On the Fabric of the Human Body)*, which provided the first accurate description of the human body based on dissections. For those with unblinkered vision, it was now clear that the Earth was just a planet and that human beings were just animals. Alas, for many people the blinkers concerning humankind's place in nature remained on for centuries to come; but it was a start.

The first steps towards an understanding that evolution has happened came from the investigation of fossils, the remains of once-living creatures preserved in ancient rocks. But that simple statement needs unpacking. First, fossils had to be recognised as the remains of living creatures; second, the rocks had to be recognised as ancient. Neither proposition was widely accepted at the beginning of the seventeenth century. Of course, many people had noticed the existence of fossils. Leonardo da Vinci (1452 to 1519) was one of the thinkers who puzzled over them. One of the key puzzles was that patterns in the rock resembling seashells were found high in the mountains, far from any ocean. In Leonardo's day the received wisdom was that these patterns, often resembling other living forms, not just seashells, were no more than an imitation of living things, perhaps formed when the rocks themselves formed, or perhaps still being formed today by some mysterious influence of the stars or the Moon. Leonardo would have none of it. He didn't know how fossils formed, but he was sure they were not supernatural. Early in the sixteenth century, he wrote in one of his notebooks:

> *If you should say that these shells have been and still constantly are being created in such places as these by the*

nature of the locality or by potency of the heavens . . .
such an opinion cannot exist in brains possessed with any
extensive powers of reasoning.

A brain possessed with plenty of powers of reasoning got to
grips with the puzzle a century and a half later.

Robert Hooke has sometimes been described as 'London's
Leonardo'. Like Leonardo, he was a polymath; he made major
contributions to astronomy and microscopy, was Christopher
Wren's architectural partner in the rebuilding of London after
the Great Fire of 1666 (many 'Wren' churches are his work),
and he was the first science populariser, notably with his book
Micrographia, described by Samuel Pepys as 'the most ingen-
ious book that ever I read in my life'. But here we will focus
on his work in the biological and Earth sciences, which was
largely ignored in his lifetime.

Hooke was born in 1635, at Freshwater, on the Isle of
Wight. The location is significant because the chalk strata of
the island, exposed in high cliffs, are rich in seashells, even
in layers high on the cliffs, far above the waves. Hooke later
remembered his childhood curiosity being fired by seeing a
layer of sand far above the sea, 'Filled with a great variety of
Shells, such as Oysters, Limpits, and several sorts of
Periwinkles'.[5] The standard explanation was that this had
something to do with the biblical flood, although exactly
what process might have been involved remained vague.

Hooke's father, the curate of All Saints' Church, would
have accepted the biblical version of events. He was responsible for the early education of Robert, who was felt to be
too delicate to be sent away to school as his elder brother
had been. While Robert was growing up, the turmoil of civil
war ravaged much of mainland England, but the Isle of Wight
emerged unscathed. When Hooke's father died in 1648,
Robert, aged thirteen, left for London with a modest legacy

that enabled him to enrol at Westminster School, where he shone at his studies, especially in mathematics. With the execution of Charles I in January 1649, order was restored under Parliament, and it was under this settled regime that Hooke went up to Oxford University in 1653, when he was eighteen. But he never took the BA examination, instead becoming an assistant to a group of gentlemen 'philosophers', including professors at the university, who were interested in what we now call science. He gleaned knowledge from these men, rather than attending formal lectures, and became more than the assistant to the greatest scientist among them, Robert Boyle (1627 to 1691). Hooke was, in effect, Boyle's partner in a series of experimental investigations. It was through these connections that when the Royal Society was founded in London in 1661, following the Restoration of Charles II at the end of the Parliamentary interregnum, Hooke became their Curator of Experiments. He quickly became the person who made the Society work, demonstrating experiments to its Fellows (many of them the same gentlemen he had worked for in Oxford) at their regular meetings, and carrying out experiments of his own. Although he had wide-ranging interests, in the early days these studies largely involved the newly invented microscope. What he referred to as his 'first endeavours' led to the publication of his great book, *Micrographia*, at the beginning of 1665. In the preface, he clearly nails his colours to the mast of the mechanistic interpretation of nature, with no mention of gods or mysterious spirits:

> *We may perhaps be inabled to discern all the secret workings of Nature, almost in the same manner as we do those that are the productions of Art [i.e. artifice], and are managed by Wheels, and Engines, and Springs, that were devised by humane Wit.*

By the time *Micrographia* was published, Hooke had examined various fossils and petrified wood and had concluded that these stony remains had indeed once been living, but that after death they had been:

> . . . *fill'd with some kind of Mudd or Clay, or petrifying Water, or some other substance, which in tract of time has been settled together and hardened in those shelly moulds.*

He later elaborated on the theme in a series of lectures he gave at the Royal Society on 'earthquakes', a term that he used to cover all kinds of changes on the surface of the Earth.[6] He stated unequivocally that fossils represent either organic matter itself turned into stone, or are impressions of living things, and was as dismissive as Leonardo of anyone who thought otherwise. The idea that they were formed 'from some extraordinary Celestial Influence, and that the Aspects and Positions of the fix'd Stars and Planets conduc'd to their Generations,' he says, is 'fantastical and groundless'.

Around the same time (but, probably significantly, after the publication of *Micrographia*), the Danish scientist Niels Stensen, usually known by the Latinised version of his name, Steno, also realised that fossils were the remains of living creatures. He had been born in 1638 and had qualified as a physician by the time he published his only important scientific work, in 1669. Its title was 'Predecessor of a dissertation of a solid naturally contained within a solid'. The solids inside solids were fossils, and he drew particular attention to examples known as tongue-stones, which he identified (correctly) as fossilised sharks' teeth. He reasoned that the rocks in which they were found must have been laid down underwater, and that since there are many such layers there must have been a series of great deluges, the latest of which could be identified with the biblical flood.

Steno's ideas gained attention in England because his work

was translated and promoted by Henry Oldenburg, the Secretary of the Royal Society. Oldenburg was no friend of Hooke, and had passed on the ideas from Hooke's early lectures on earthquakes to Steno. Even if this information did no more than reinforce Steno's views on fossils, Oldenburg's active promotion of the Dane's work helped to obscure the significance of Hooke's earlier, and more complete, contribution. Steno was never in a position to respond to any criticism Hooke may have made, because he gave up science (the 'dissertation' referred to in the title of his book never appeared), became a Catholic priest with extreme ascetic inclinations, and died at the age of 48, partly because of his severe regime of fasting and self-denial.

Hooke, though, went much further than Steno and had a much deeper insight into the reason why marine fossils were found so far from the sea and so high above sea level. After describing how fossils are found at the tops of the highest hills, in the depths of the deepest mines, and in stone quarries in mountains far from the sea, he explains that this can only have happened if the surface of the Earth had been 'transform'd, and made of another Nature' as time passed. 'Parts which have been Sea are now Land, and others that have been Land are now Sea; many of the Mountains have been Vales, and the Vales Mountains.'

And he elaborates on what he means by a 'tract of time':

Nor do I conceive they were all thus formed at once, but rather successively, some in one, some in other Ages of the World, which may probably be in some measure collected from the quantity or thickness of the Soil or Mould upon them fit for Vegetation.

He not only realises that the Earth must be much older than the few thousand years accepted by biblical scholars of his day, but that successive layers of rock might be dated by measuring their depth beneath the surface. We shall look at the influence of Hooke's insights into geology and the age of the Earth in the next chapter, but his ideas about evolution were, if anything, even more profound.

Hooke's 'earthquake' lectures were given at various times over the last four decades of the seventeenth century, and after his death they were collected and published by his friend Richard Waller as *The Posthumous Works of Robert Hooke.** This book appeared in 1705, a century and a half before *On the Origin of Species*, and seems to have made no impact at all as far as its revolutionary ideas about life are concerned. Hooke recognised that if fossil ammonites were the remains of living creatures, and there were no ammonites around today, it meant that species could become extinct. And this suggested to him that new species could emerge in the course of time:

> *There have been many other species of Creature in former Ages, of which we can find none at present, and that 'tis not unlikely also but there may be divers new kinds now, which have not been from the beginning.*

And:

> *Since we find that there are some kinds of Animals and Vegetables peculiar to certain places, and not to be found elsewhere; if such a place have been swallowed up, 'tis not improbable that those animal Beings may have been destroyed with them.*

How could he explain the origin of these new species? By environmental change:

---

* The modern transcription by Drake is the best place to find them today.

*There may have been divers new varieties generated of
the same Species, and that by the change of the Soil in
which it was produced, for since we find from the alteration of the Climate, Soil and Nourishment doth often
produce a very great alteration in those Bodies that
suffer it.*

His overall conclusion is, if not quite Darwinian, certainly
remarkable for someone born two hundred years before the
Beagle and its naturalist returned home from their voyage
around the world:

*Certainly, there are many Species of Nature that we have
never seen, and there may have been also many such
Species in former Ages of the World that may not be in
being at present, and many variations of those Species
now, which may not have had a Being in former times
. . . it seems very absurd to conclude, that from the beginning things have continued in the same state that we now
find them.*

All this was published in 1705. Hooke understood that the
Earth had a very long history, that there had been what we
now call mass extinctions of life, and that new species had
emerged after those extinctions. But it was a false dawn.
Entirely unaware of Hooke's contribution, eighteenth-century
scientists independently worked their way towards an understanding of evolution.

Before they could develop a comprehensive theory of
evolution, they needed a clear understanding of species and
their relationship to one another. The first detailed description
of this kind came from the Swedish botanist Carl Linnaeus
in the 1750s, but he drew on the earlier work of John Ray,
a slightly older contemporary of Robert Hooke.

Ray came from a modest but not impoverished background. Born in 1627, he was the son of an Essex blacksmith and a local herbalist (a kind of 'wise woman' who treated sick villagers); both parents were important members of their small community. Ray's ability at school was noticed by the rector, who arranged for him to go on to the grammar school in Braintree, where the vicar, a graduate of Trinity College Cambridge, took Ray under his wing and arranged for him to go up to Cambridge in 1644. He was only able to do so because he was admitted as a 'subsizar', a student who paid his way by acting as a servant for gentleman scholars. Although he was expected to take holy orders and become a priest after he graduated in 1648, the situation was complicated by a religious wrangle between the university and Parliament, which had abolished bishops, and Ray was not ordained, although he became a Fellow of Trinity College. He spent the years up to 1660 reasonably successfully in several teaching posts, doing well enough to provide a house for his mother in her home village when she was widowed in 1655. As a Fellow of Trinity, he had comfortable accommodation and freedom to follow his own interests, which increasingly turned to classifying the similarities and differences between plants, with the aid of any interested students. But with the political changes at the end of the 1650s and the Restoration, all the old Church rituals, including bishops, were brought back and Ray was ordained, fully intending to become a parish priest. Then came a twist; Parliament had abolished bishops as part of a law requiring all clerics to swear an oath known as the Covenant, part of a general reorganisation of the Church. Charles II now ordered all clerics to declare formally that this act had been unlawful and their oaths were null and void. Although Ray had never 'taken the Covenant' himself, he believed that an oath was a commitment before God and therefore could not be broken or revoked, so he refused to make this declaration. Anticipating the reaction of the authorities, he resigned all his posts and became an unemployed priest. As a priest, he could not take secular work, but he could not practise as a priest either because of his stand against oath breakers and a king who incited people to break oaths.

He was saved from this dilemma by a wealthy Cambridge friend, Francis Willughby, who had been one of his group of plant collectors. Willughby took him on a trip round Europe to study the animal and plant life. They set off in April 1663, and Ray did not return until the spring of 1666, his mind and notebooks full of information about the living world, and armed with many specimens to pore over back at home. He became a Fellow of the Royal Society the following year, and went on other expeditions around England. He became part of the Willughby household, but after Willughby died in 1672, Ray married and eventually returned to Essex, living modestly off the rent he received from some land owned by his family. He now had ample time to work on his epic *History of Plants*, which appeared in three volumes – the third published in 1704, the year before Ray died.

Ray didn't only write about plants. He had previously written books on fishes and birds (published under Willughby's name, but largely Ray's work), and after his death a posthumous volume on 'insects' appeared – in those days, 'insect' was a catch-all term for anything that wasn't a bird, animal or fish. As well as gathering together a wealth of material in an accessible form, he devised a taxonomical system that classified species in terms of their physiology, anatomy and morphology. This was the first systematic classification system, which made the study of botany and zoology properly scientific. In spite of his deeply held religious beliefs, Ray also puzzled over the significance of fossils, and he recognised the difficulty of reconciling observations of the real world with a literal interpretation of the Bible.

In 1663, after observing the remains of a buried forest near Bruges, he wrote:

> *Many years ago before all records of antiquity these places were part of the firm land and covered with wood; afterwards being overwhelmed by the violence of the sea they continued so long under water till the rivers brought down earth and mud enough to cover the trees, fill up these shallows and restore them to firm land again . . . that of old time the bottom of the sea lay so deep and that hundred-foot thickness of earth arose from the sediment of those great rivers which there emptied themselves into the sea . . . is a strange thing considering the novity of the world, the age whereof, according to the usual account, is not yet 5600 years.[7]*

Although his classification system went further, Linnaeus leaned heavily on Ray's pioneering work, but, always eager to polish his image (he wrote five self-serving autobiographical memoirs), he avoided giving this source due credit. This is particularly unfortunate because his own achievements were so great that the image of Linnaeus hardly needed polishing, and his efforts to do so actually tarnish it a little.

Linnaeus was born in 1707, and his clergyman father intended that the boy should follow in his footsteps. But as Carl showed no inclination or aptitude for this, let alone a vocation, he was allowed to study medicine, first at the University of Lund, then, from 1728, in Uppsala.

While he was still a student, Linnaeus became intrigued by the then new idea, put forward in 1717 by the French botanist Sébastien Vaillant, that plants reproduced sexually. Vaillant identified male and female parts in plants, but nobody at the time (including Linnaeus) properly understood the part played by insects in the process of pollination. Linnaeus, whose father was a keen amateur botanist, had been fascinated by flowering plants since he was a child, and now came up with the idea of using the reproductive parts as a means of identifying and classifying plants. In 1729 he wrote a thesis on plant sexual reproduction, which led to him giving lectures and demonstrations at the botanical gardens in Uppsala, as a stand-in for the professor, Olof Rudbeck, when he was still a second-year student. Rudbeck had been on a botanical expedition to Lapland in 1695, but his notes and specimens were lost in a fire in 1702; under Rudbeck's influence, Linnaeus went on a similar expedition in 1732, financed by the Royal Society of Sciences in Uppsala. By this time he was already developing a classification system for plants based on the number of stamens and pistils in their flowers. While he was on this expedition, he came across the jawbone of a horse by the roadside, and according to his later recollection he realised that 'if I only knew how many teeth and of what kind every animal had, how many teats and where they were placed, I should perhaps be able to work out a perfectly natural system for the arrangement of all quadrupeds'.[8] This was a natural leap for Linnaeus, an obsessive list-maker and classifier who did everything by the clock or the book (his book). What would have been a handicap in many circumstances made him the ideal person for his chosen life's work.

By the time Linnaeus moved on to the Netherlands to complete the academic requirements for his medical qualification – he already had a nearly finished thesis to present to the University of Harderwijk to earn his doctorate, awarded in 1735 – he had completed his first attempt at a classification of plants. This was published in the Netherlands as *Systema Naturae* in the same year that he completed his medical studies. He stayed in the Netherlands until 1738, working as a physician, then visited England in July 1736, meeting botanists in London and Oxford and botanical colleagues in Paris on the way back to Sweden. He was back in Sweden in June 1738 and never left the country again. He married Sara Moraea, the daughter of a doctor, the following year, and practised medicine in Stockholm until 1741. He then became Professor of Medicine in Uppsala, but switched to the Chair of Botany in 1742, and stayed in that post until his death, in 1778.

As professor of botany, Linnaeus was able to give full rein to his nerdish organisational skills. He would take students out on botanical day trips timetabled to the minute. The party, dressed in special light clothing of his devising, always set out promptly at 7 a.m., with Linnaeus giving a demonstration every half hour on the dot. Lunch was taken at 2 p.m., and there was a short rest at 4 p.m. This obsessive attention to detail showed in his published work. During his time as professor of botany, Linnaeus was constantly writing new books, and revising his *Systema Naturae*. The idea of classifying each species using a two-word name ('borrowed' from Ray but extended by Linnaeus) first appeared in his book *Species Plantarum* in 1753, and then in the tenth edition of the *Systema Naturae*, in 1758. Using his own studies in the field, and drawing on the work of predecessors such as Ray, Linnaeus published in various places descriptions of more than 7,500 species of plants and 4,400 species of animals. Each was given a unique binomial term, specifying its genus and species – for example, *Canis lupus*, the wolf. Although the list has been extended and modified over the centuries, it is thanks to Linnaeus that a biologist can refer to a species by name – such as *Canis lupus* – and be sure that any other biologist will know exactly which animal or plant is being referred to. The classification extends upwards – from species to genus, family, order, class and kingdom. In volume ten of the *Systema*, Linnaeus also introduced many terms, including

Mammalia, Primate and *Homo sapiens*, making it possible to locate our own place in the biological world. Slightly updated to modern terminology, it looks like this:

Kingdom: Animalia
Phylum: Chordata
Subphylum: Vertebrata
Class: Mammalia
Order: Primates
Family: Hominidae
Genus: Homo
Species: sapiens

Linnaeus agonised over this classification of human beings. In the mid-eighteenth century it was daring even to suggest that people could be classified in the same way as animals. Yet as Linnaeus wrote in his *Fauna Svecica (Fauna of Sweden)*, published in 1746, 'I have yet to find any characteristics which enable man to be distinguished on scientific principles from an ape'. The following year, in a letter to a colleague he wrote:

> *I ask you and the whole world for a generic differentia between man and ape which conforms to the principles of natural history. I certainly know of none . . . If I were to call man ape or vice versa, I should bring down all theologians on my head. But perhaps I should still do it according to the rules of science.*[9]

In the end, to avoid the wrath of theologians and by bending the rules of science, Linnaeus compromised by placing our own species in a genus of its own Homo. In modern times, other (extinct) species of *Homo* have been included in the genus, but modern studies of DNA confirm the evidence of outward appearance; by any reasonable scientific classification scheme we should be included with the chimpanzee in the genus *Pan*, with the gorilla as an almost equally close relation. But what is important here is that Linnaeus knew this more than a century before the publication of the *Origin*, and 140 years before Darwin dared to challenge the theologians by publishing *The Descent of Man*. Nevertheless, Linnaeus was religious and did not think that new species could evolve, even though he recognised that new varieties of plants came into existence from time to time. Ironically, the term 'evolution' was introduced to biology by another man who believed in the fixity of species and the role of God in creating them, a contemporary of Linnaeus, Charles Bonnet.

The key feature of a species, of course, is that members of the same species can reproduce with each other and produce offspring that can also breed with other members of the species. Male and female horses, for example, mate to produce horses; male and female donkeys mate to produce donkeys. But even though a male donkey and a female horse can mate to produce offspring – a mule – the offspring are sterile, because horses and donkeys are different species. In the world of plants and animals, as in this example, reproduction like this usually involves sex, but Bonnet was astonished to find proof that one species, at least, could reproduce without the benefit of sex.

Bonnet was born in Geneva, then still an independent republic, in 1720, and seems to have stayed there for his entire life; he died in 1793. He studied law, in accordance with his father's wishes, and practised it after a fashion, but his family was wealthy enough for him to indulge his real interest, the study of the natural world. His many observations included noticing that bubbles form on plant leaves that are immersed in water, showing that plants are releasing gas, and the discovery that caterpillars and butterflies breathe through pores, which he named stigmata. But his most dramatic discovery was that female aphids could produce young without any contribution from a male – what is now known as parthenogenesis. Bonnet's interest in insect reproduction was partly inspired by correspondence with his uncle, Abraham Trembley (1710 to 1784), who was working as a tutor to a rich family in the Netherlands. Trembley would shortly become famous for experiments on the tiny water creatures known as hydra, which seemed to be intermediate between plants and animals – they could move, like animals, but when cut in two each part would regenerate into a whole new creature, just like a cutting from a plant. But in 1740, when Bonnet was still a law student, one topic of the correspondence concerned the nature of aphids. Trembley and his contemporaries had been unable to find a single male aphid, yet the little insects undoubtedly reproduced, with females giving birth to young, known as nymphs. Bonnet decided to solve the mystery. He put a single newborn nymph on a branch of shrubbery inside a sealed glass container, and watched over it in his room from 20 May to 24 June, making sure that the vessel was not disturbed. The female aphid gave birth to her first daughter on 1 June, and between then and 24 June produced another ninety-four offspring. Within weeks, the observations of 'virgin birth' had been confirmed by other researchers, including Trembley, and at the age of twenty Bonnet was appointed a Corresponding Member of the French Academy of Sciences. In further experiments, he raised thirty generations of virgin aphids that never encountered a male.

But what did it all mean? Even though Bonnet discovered a male aphid in December 1740, the fact remained that unmated virgin aphids could reproduce. Bonnet had ample opportunity to ponder the implications, because his eyesight failed and, unable to carry out experiments any more, he turned his attention to philosophical issues, which he discussed in a series of widely read books. His explanation of parthenogenesis was that every individual aphid had already been created by God in the beginning, and that they were nested inside one another (like Russian dolls), ready to emerge at their proper time.

This idea, known as preformation, was popular at the time. The behaviour of the parents, or environmental influences, might affect how individuals developed once they were born, but the basic individual was entirely God's design. This did not only apply to aphids, but to all species, where the role of the male was only seen as some kind of trigger to stimulate the growth of the next individual inside its mother. It was in this context that Bonnet introduced the term 'evolution' to biology in 1762, in his book *Considerations sur les corps organisées (Considerations on Organised Bodies)*. The word comes from the Latin *evolutionem*, meaning unrolling – as in the unrolling of a scroll to reveal what has already been written (in this case, by God). This is the exact opposite of its modern meaning, which is why Charles Darwin was famously reluctant to use the term (it does not appear at all in the *Origin*, and he preferred the phrase 'descent with modification'), although his grandfather, as we shall see, had no such scruples. But even in Bonnet's time, it was clear that Bonnet's version of preformation was wrong. As early as 1745, in his book *Earthly Venus*, Pierre-Louis Moreau de Maupertuis (1698 to 1759) had summed up the evidence that an embryo does not start out as a tiny version of the adult and simply get bigger, but develops by epigenesis, with different features appearing one after the other.

Maupertuis was another member of a wealthy family with no need to earn a living, although he led a rather more exciting life than Bonnet. He was born in Saint-Malo, educated privately, and became an officer in the cavalry, which was a largely honorific post that gave him plenty of time to socialise with the gentry and indulge his interest in mathematics. After leaving the cavalry and moving to Paris, in 1723 he became a member of the Academy of Sciences and was an early Continental supporter of the ideas of Isaac Newton, which had initially been viewed with suspicion in France as the work of a mere Englishman. Maupertuis is best known in science for his work in physics and mathematics; he was the head of a French expedition to Lapland in 1736 to measure the length of a degree of arc of the Earth's circumference, and proposed an idea known as the principle of least action, which in essence says that nature follows the cheapest option (the fact that light travels in straight lines is an example), although he did not put it on a secure mathematical footing.

Maupertuis also became a real soldier. In 1740 he was invited to Berlin by the Prussian King, Frederick II. On the outbreak of a war between Prussia and Austria he offered his military services, and was captured by the Austrians at the Battle of Mollwitz in 1741. When he was released, he went briefly to Berlin and then back to Paris, where he became the Director of the Academy of Sciences in 1742. Two years later he was again headhunted by Frederick II, and in 1746 he became President of the Royal Prussian Academy of Sciences. But his position became uncomfortable with the outbreak, in 1756, of the Seven Years' War, in which France and Prussia were on opposite sides. He became *persona non grata* in Berlin because he was French, but he was regarded with suspicion in France because of his close ties with Frederick II. He retired to the south of France, and then to Basel, where he died, but he found time during these upheavals to write a book, *Venus physique (The Earthly Venus)*, which was published in 1745, and in which he expounded his ideas about evolution.

These ideas were not always clearly expressed, and the work suffers from his espousal of the idea of grotesque forms arising by chance before being selected. However, Maupertuis does endorse selection in his book:

Chance, one would say, produced an innumerable multitude of individuals; a small number found themselves constructed in such a manner that the parts of the animal were able to satisfy its needs; in another infinitely greater number, there was neither fitness nor order: all of these latter have perished. Animals lacking a mouth could not live; others lacking reproductive organs could not perpetuate themselves.[10]

This was no more than others had said, going back to the Ancient Greeks, but Maupertuis did make a significant contribution to the understanding of heredity, in the process pulling the rug from under the idea of preformation. He realised that an embryo is formed from a combination of material from both parents, and develops from this combined seed. He was especially interested in the occurrence of polydactyly (the presence of an extra finger) in people. The preformation idea would have it that this abnormality was built in by the Creator in the beginning, ready to be 'unrolled' when the time came. Maupertuis said that it was an accident, and noted that the abnormality could be passed on to future generations by either parent. Both parents could pass on features to their offspring, so how could all the generations back to Eve be preformed inside their mothers?

During his lifetime, these ideas were expressed most clearly in letters to colleagues; but they were included in a posthumous collection of his works, published in 1768. Maupertuis went astray, however, in suggesting in *The Earthly Venus* that changes in the bodies of the parents could affect the material from which the 'seeds' produced by the parents formed. He concluded that the changes were passed on to their offspring, even to the point where animals could spontaneously develop new organs in response to outside influences, organs which would then be inherited by their descendants. A less extreme version of this idea would be developed by Jean-Baptiste Lamarck early in the next century.

The Earthly Venus also had a big influence on one of Maupertuis' contemporaries, Denis Diderot, a French free-thinker and major figure in the Enlightenment, who had been born in Langres, in the Champagne region, in 1713, and had been living in Paris as a dropout and writer after abandoning his study of law in 1734 and being cut off by his father. This set the tone for a bohemian hand-to-mouth life in which he was often in trouble with the authorities and even imprisoned for six months (July to December 1749) for his anti-establishment writings. His life's work was an encyclopedia, published in volumes as time passed, in which he offered knowledge to the people to encourage them to think for themselves. The first volume appeared in 1751, and the project was almost immediately seen as seditious by the authorities, who feared its influence on the masses; in 1759 it was formally suppressed, but work continued undercover with great difficulty. The project was only completed in 1772. The following year Diderot visited Russia at the invitation of Catherine the Great; he stayed for five months and she was sufficiently impressed to give him 3,000 roubles to cover his expenses (twice what he had asked for), and a valuable ring. In 1784, when she heard that he was ill, Catherine arranged for him to be moved into comfortable accommodation, where he died a few weeks later.

The encyclopedia was by no means Diderot's only work, but here we are only interested in his thoughts on evolution. His insight can be seen in one sentence from the encyclopedia: 'Nature advances by nuanced and often imperceptible degrees.' He saw that, rather than producing monstrous varieties of individuals from which to select, evolution proceeds in tiny steps. This was a major advance in the second half of the eighteenth century. Expanding on the theme, he wrote:

> *May it not be that, just as an individual organism in the animal or vegetable kingdom comes into being, grows, reaches maturity, perishes and disappears from view, so whole species may pass through similar stages? If the faith had not taught us that animals came from the hands of the Creator just as they are now, and if it were permissible to have the least uncertainty about their beginning and their end, might not the philosopher, left to his own conjectures, suspect that the animal world has from eternity had its separate elements confusedly scattered through the mass of matter; that it finally came about that these elements united – simply because it was possible for them to unite . . . that millions of years have elapsed between each of these developments; that there are perhaps still new developments to take place which are as yet unknown to us . . . But religion spares us many wanderings and much labour. If it had not enlightened us on the origin of the world and the universal system of beings, how many different hypotheses would we not have been tempted to take for nature's secret?*[11]

Diderot was an atheist, and he had his tongue firmly in his cheek when expressing these thoughts. It is obvious why he was feared by the authorities in Catholic France in the decades leading up to the Revolution.

But by then even some religious believers were beginning to grasp the truth about evolution. James Burnett, who became known as Lord Monboddo, was born at his father's estate, Monboddo House, Kincardineshire (on the northeast coast of Scotland), in 1714. He studied in Aberdeen, Edinburgh and Groningen, qualifying in law and eventually (in 1767) becoming a judge in Edinburgh. But he was also a philosopher strongly influenced by Aristotle, and was especially interested in the origins of language. It was here that what we would now call his scientific insight came up against his strongly held religious views. The result gives us an almost perfect example of how difficult it was to establish a complete theory of evolution until the religious shackles had been cast off.

Monboddo, as he is usually referred to, studied languages from widely separated parts of the world, including those of Native Americans and Tahitians, as well as those of Northern Europe and the Middle East. He developed the idea that languages had evolved, and this led him to suggest that human beings had emerged in one location and spread across the Earth. This was the first scientific proposal of the single origin idea of humanity, although, of course, it was entirely in line with the story of Adam and Eve. But Monboddo went further; he saw that humans were related to primates, and sometimes referred to apes as our 'brothers'. His hypothesis of the origin of language involved physical changes in the speech organs over many generations, as people adapted skills to cope better with their environment, starting out from some ancestral form even more like that of our brother apes. He bred horses, and was well aware of the possibility of changing the form of a species by selective choice of mates to breed from – for example, always breeding from the largest horses to produce bigger, stronger individuals. This artificial selection was a procedure that Charles Darwin himself used as a jumping-off point for his ideas. And Monboddo even came up with an overarching model of the development of modern people, with tool use coming first, then the appearance of social structures, and finally language.[12]

But how could all this be squared with the biblical story of the Creation? Monboddo was quite happy with the idea that those stories were allegories and not to be taken literally, but he was equally happy with the idea that the Universe had indeed been created by God. In order to square the circle, in the 1770s, in his multi-volume *Of the Origin and Progress of Language*, Monboddo argued that humans had descended from apes, but that apes themselves are to be grouped with humans as a distinct creation, separate from the rest of the animal kingdom.

Although Monboddo's ideas were not a major influence on the generations that followed, he was known to evolutionary thinkers such as Erasmus Darwin, and even to a wider audience – in Charles Dickens's *Martin Chuzzlewit*, there is a mention of 'the Monboddo doctrine touching the probability of the human race having once been monkeys'. This appeared in print in 1843, sixteen years before *On the Origin of Species* was published, and 44 years after Monboddo's death.

It's a sign of the confusion surrounding the development of evolutionary ideas in the second half of the eighteenth century that Monboddo's contemporary, Georges-Louis Leclerc, the Comte de Buffon, came up with the first scientific estimate of the age of the Earth and, without reference to God, saw evolution at work in the living world, but could not accept the idea that humans and apes shared a common descent, and debated the issue with Monboddo in their correspondence. If only their ideas had been combined, there might have been at least a small leap forward before 1800.

With Hooke's ideas having been overlooked, Buffon's contribution marks the beginning of the genuinely scientific investigation of the origin of the Earth and the evolution of life on Earth, an investigation that extends from his time in an unbroken line to Charles Darwin and beyond, although there were many mistaken branches from that line. Although he started life, in 1707, as Georges-Louis Leclerc, we shall refer to him throughout as Buffon, to avoid confusion. He came from Montbard, near Dijon, where his father was an official involved in collecting the salt tax. The name Georges was carefully chosen; Buffon's mother's uncle, Georges Blaisot, was a much more wealthy tax 'farmer', who had no children of his own and became Buffon's godfather. When he died in 1714, he left a large fortune for the benefit of the boy, although in his infancy it was administered by his parents. Buffon's father, Benjamin François Leclerc, interpreted his role in this liberally; he bought large amounts of land, including the entire village of Buffon, moved the family to Dijon, and became a councillor in the local parliament. Georges-Louis went to a Jesuit college in Dijon, then studied law before moving to Angers to study mathematics, botany and medicine; along the way he also seems to have studied astronomy. But in Angers he met the young English Duke of Kingston, who was on the Grand Tour of Europe, and in 1730 he gave up his studies to travel with him. It was during this tour that he added 'de Buffon' to his name, not wishing to be too overshadowed by his friend's title. The Duke certainly travelled in grand style, with servants and carriages, and staying in magnificent lodgings. Buffon found the life easy to get used to, and he soon had the means to follow it.

In August 1731 Buffon's mother died, and in December the following year his father remarried and tried to appropriate the entire family fortune. After legal wrangling, Buffon secured his inheritance, including the village of Buffon, and a substantial fortune, if not quite on the scale of that of the Duke of Kingston. He pointedly dropped the paternal 'Leclerc' from his name, and started styling himself 'Georges-Louis de Buffon' and signing his name simply as 'Buffon', as if he were also a member of the nobility. In August 1732 he had settled in Paris, where he could have spent the rest of his life as a member of the idle rich. Instead, he mixed with intellectuals, including Voltaire, and studied science, as well as being an active and forward-thinking manager of his estates in Burgundy. His prodigious scientific output over the rest of his life is partly explained by a working regime designed to overcome what he thought was his natural laziness. A servant was employed to wake Buffon at 5 a.m. and if necessary physically drag him out of bed. Work started immediately, with a break for breakfast (two glasses of wine and a bread roll) at 9 a.m., then work until 2 p.m., with the afternoon set aside for lunch and entertaining any visitors, followed by a short nap and a long walk, then work from 5 p.m. to 7 p.m., no supper and in bed by 9 p.m.

Buffon first made his mark in mathematics, in particular probability theory, where a problem known as Buffon's needle is named after him.* In 1734 he became a member of the French Academy of Sciences, for whom he carried out a study of the structural properties of wood, which was very important in those days, when naval power depended on wooden ships. In June 1739, at the young age of thirty-one, Buffon became an Associate Member of the Academy. A month later, the superintendent of the French botanical garden, the Jardin du Roi, died unexpectedly and Buffon was the right man in the right place (and with the right contacts) to be given the post. Not least of his qualifications for the role was that he did not need to draw his salary; the Jardin was essentially bankrupt, and he was among the patrons who occasionally provided funds to keep it going. It was almost a bonus that he turned out to be very good at the job, which he held for the next forty-one years, turning the Jardin into a major centre of research, extending its grounds, and acquiring botanical and zoological specimens from many regions of the globe.

Buffon's masterwork was his attempt to cover the whole history of the natural world, the *Histoire naturelle*, which

---

* If we have a floor made of parallel strips of wood, each the same width, and we drop a needle of a certain length onto the floor, what is the probability that the needle will lie across a line between two strips?

appeared in forty-four volumes between 1749 and 1804 – the last eight published after his death (Buffon's brother, incidentally, was a contributor to Diderot's encyclopedia). The *Histoire* was influential not just because it was so comprehensive, but because Buffon wrote in a clear style that appealed to the general reader and made it a best-seller; it was said to have been essential reading for every educated person in Europe. He was, indeed, such a fine writer that he was appointed to the Académie Française in 1753. This was the year after he had married; his wife died in 1769, five years after she had borne a son who proved to be just the kind of rich wastrel Buffon might have been, and who (to put it mildly) lacked his father's intellect; he ended up as a victim of the Terror after the French Revolution of 1789. Buffon himself died in 1788, so he missed all that excitement, but he had been awarded the title of Comte in 1772, at last justifying his habit of styling himself 'Buffon'.

The *Histoire* began with three volumes published in 1749, in which Buffon outlined his model of the origin of the Earth, which did not even pay lip service to the biblical story, and included his estimate of the age of our planet. Buffon picked up on a suggestion made by Isaac Newton that the Earth had been formed from material torn out of the Sun by the impact of a comet. At the time, it was thought that the Sun was a glowing ball of hot iron, and in his *Principia* Newton had suggested that a ball of red-hot iron the size of the Earth would take at least 50,000 years to cool to its present state. This was a daring extension of the age of the Earth calculated from biblical chronology by Archbishop James Ussher and published in 1650, which suggested that the Creation occurred in 4004 BCE, but does not seem to have provoked much response in Newton's day. Newton himself did not try to make a precise measurement of how quickly hot balls of iron cool, but said, 'I should be glad that the true ratio was

investigated by experiments.' It was Buffon who took up the
challenge.

Buffon took balls of iron of different sizes, heated them
until they were on the point of melting, then measured how
long they took to cool down. There were no accurate ther-
mometers in those days, so he persuaded aristocratic ladies
with delicate hands to take part in the experiments. Wearing
the finest silk gloves, they were asked to judge the point
where the iron balls were just cool enough to be held in their
hands without burning. As expected, the larger globes took
longer to cool. By extrapolating from his measurements to a
ball the size of the Earth, Buffon calculated that it would
have taken the planet 75,000 years to cool to the same point,
and included this estimate in the *Histoire*, although he realised
that the actual age of the Earth must be even greater. He
wrote, for public consumption:

*I aimed at determining two moments during cooling:
the first one when the balls stopped burning anymore,
i.e. when one could touch them and hold them in one's
hand during a second without being burned. The second
one when the balls were cooled down to room temperature, i.e. ten degrees above the freezing temperature. To
determine the time when the ball reached room temperature, we compared it with other cannonballs with the
same diameter, but which had not been heated and could
be touched at the same time as those that had been
heated. Through this simultaneous and instantaneous
touching on two balls with one hand or both hands, we
could determine the moment when both balls were
equally cold . . .*

*Now if we wanted to infer with Newton how much
time was needed for a sphere as big as the Earth to cool
down, one would find according to the above experiments*

that instead of the fifty thousand years he had estimated for the Earth's cooling time to be down to its present-day temperature, one needed forty-two thousand nine hundred sixty-four years and two hundred twenty-one days to cool it down to a temperature where it would not burn, and ninety-six thousand and six hundred seventy years, and one hundred thirty-two days to cool it to room temperature . . .

Assuming, as all the phenomena seem to indicate, that the Earth had been once liquid because of the fire, our experiments demonstrated that if the sphere had been completely composed of iron or ferruginous matters it would have solidified down to its core only in 4026 years, cooled to be touched without burning the fingers in 46,991 years only, and reached room temperature only in 100,696 years; however, as the Earth, according to all we know, seems to be made up of vitrifiable matters and limestones that cool faster than the ferruginous ones, one must consider, in order to get near the truth as much as possible, the respective cooling times of the various materials as we measured them through our experiments in our second Mémoire, and infer the ratio with the iron cooling time. By using in this sum only glass, sandstone, hard limestone, marble, and the ferruginous matter, one finds that the Earth sphere solidified down to its centre in about 2905 years, that it cooled enough to be touched in c. 33,911 years, and to room temperature in c. 74,047 years.

But his manuscripts show that he actually thought that the Earth might be as much as three million years old, because he realised that the deposition of the sediments that form the rocks of the Earth's surface was a process that would have taken an immense time. These figures remained unpublished, probably because he knew the disbelieving reaction that they would provoke. For various reasons (including the fact that the crust of the Earth acts as an insulating blanket trapping heat inside), even this was a huge underestimate; but it was still the first scientific attempt to calculate the age of the Earth. Another Frenchman, Joseph Fourier, was able to improve on Buffon's estimate early in the nineteenth century. He developed equations to describe how heat flows from a hotter object to a cooler one, allowed for the insulating blanket effect, and made other refinements. In 1820 he published details of his technique, but he never put in print the number that came out of the calculations, and which any competent mathematician could have found by following his working. It was 100 *million* years. It seems that Fourier was, like Buffon, unwilling to risk any backlash from publishing such a figure, although he must have worked it out for himself.

Even Fourier's estimate, though, was much less than the minimum age for the Earth guessed by an earlier French thinker, whose work is sometimes overlooked.

Benoît de Maillet (1656 to 1738) was a French aristocrat and diplomat who served as Consul General in Cairo from 1692 to 1708, and took advantage of his time there, and on other overseas postings, to study the natural world and draw his own conclusions about the origin and evolution of life. His ideas were summed up in a book that took many years in the writing, going through many drafts, and was only published ten years after his death, after suffering extensive revision by an unsympathetic editor – although enough of the original manuscripts survived for de Maillet's original thoughts to have been largely reconstructed by modern historians. The book was presented as describing the philosophy of a mysterious Indian sage, Telliamed (de Maillet's name reversed) and was originally published anonymously. Although full of strange and even nonsensical ideas, including descriptions of mermaids and mermen presented as actual observations, *Telliamed* contained nuggets of scientific insight.

Like Hooke and Steno, de Maillet recognised that the presence of fossils high in the mountains implied that the rocks in which the fossils were preserved must have been laid down underwater. But unlike Hooke, he did not consider it possible that the solid rocks could have risen up over the course of time, and inferred that the Earth had once been entirely covered by sea, which had slowly receded. He did, however, recognise that this required vast tracts of time, and incorporated this idea into his discussion of the evolution of life. His mythical Indian sage Telliamed estimated that the Earth is billions of years old – ten times the age that would be calculated by Fourier in the next century.

According to de Maillet, life emerged spontaneously in the sea, having been seeded by spores from space, appearing first in the shallow waters surrounding the first mountain tops to emerge as islands from a watery world. As the water retreated, life moved onto the land, with seaweed developing into trees and shrubs, flying fish developing into birds, and fish developing (eventually) into people.

The key word here is 'eventually'. De Maillet saw that different layers of rock contained the remains of different kinds of plants and animals, including many that could not be found on Earth today. His understandably confused ideas about how one form of life could have been replaced by another are not worth going into in detail, but he did recognise that what we now call evolution must have occurred, that it took a very long time to, as he might have put it, change a fish into a man, and that life changed in response to changes in the environment.

The book provoked just the savage response that de Maillet must have anticipated. In his *Natural History*, published in 1757, the French naturalist Dezallier d'Argenville raged at the folly of bringing 'man out of the depths of the sea, and, for fear that he should descend from Adam, to give us marine monsters for ancestors!' Voltaire was equally offended, describing de Maillet as a 'charlatan' who 'wanted to imitate God, and create a world with words'.[13] This was written in 1772, when Buffon was already well on the way to describing the evolution of life on Earth without reference to God.

Although Buffon did not envisage a timescale as long as the one put forward in the *Telliamed*, he did a much better job than de Maillet of trying to explain how life had developed as conditions on Earth changed over the course of time. He shared de Maillet's view that the early Earth had been covered by water, which on his model had fallen as rain as the surface cooled, and had gradually dried out – an idea, incidentally, that carried with it the implication that the planet itself was not eternal and unchanging, but had 'evolved' as time passed. And he knew that the fossil record showed that earlier forms of life had gone extinct. But his cooling Earth idea gave a mechanism for him to attempt to explain how changes in the forms of life on Earth had occurred.

In Buffon's day, the remains of large creatures such as the mammoth had already been found at high northern latitudes. These creatures clearly resembled the elephants that are only found in warmer regions of the globe today. Buffon reasoned that when the Earth was warmer, elephant-like animals could have lived much farther north, but that as the planet cooled they had migrated towards the Equator. But he recognised that modern elephants are not identical to the fossil remains found in the north, and he studied the relationships between many species (we have only used one example for simplicity) in order to develop an idea of how they changed over time. Buffon's views also changed over time, as do the views of any good scientist when confronted with new evidence, which has sometimes confused non-scientist historians trying to work out what Buffon 'really thought' about evolution. However, the basics of his idea are quite clear once you accept that he did not hold contradictory views at the same time, but adapted his ideas as his knowledge improved.

Buffon disagreed with Linnaeus on some points. Notably, he thought that the way Linnaeus had grouped species into genera was a figment of the human imagination and the result of a desire to find patterns. At first, Buffon firmly resisted the idea of a close relationship between humans and apes. But all this was largely because initially his idea of a species was, if anything, even more extreme than that of Linnaeus. At the end of the 1740s, Buffon was referring to species as fixed entities that did not change as time passed. His view of species and their preservation was firmly grounded in his ideas about reproduction, which he recognised involved contributions from both parents. He described this as involving material from each parent mixing to become arranged into an embryo, which developed according to a pattern that he called an 'internal mould' that was unique to each species. He had no way of explaining what this mystic 'mould' might be or how it might work, except that it ensured that each generation was the same as its parents – that species did not change.

In 1753, writing in volume IV of the *Histoire*, he offered what seemed to be a reason for opposing the idea that species evolved:

> *If we once admit that there are families of plants and animals, so that the ass may be of the family of the horse, and that the one may only differ from the other through degeneration from a common ancestor, we might be driven to admit that the ape is of the family of man, that he is but a degenerate man, and that he and man have had a common ancestor, even as the ass and horse have had. It would follow then that every family, whether*

*animal or vegetable, had sprung from a single stock,
which after a succession of generations, had become higher
in the case of some of its descendants and lower in that
of others.*

The key words seem to be 'if we once admit', which echo
Diderot's 'If the faith had not taught us'. It looks as if Buffon
is offering a *reductio ad absurdum*, and saying that this prepos-
terous idea could not be true; but it is tempting to wonder
whether this is a smokescreen to hide the fact that he really
did already accept the idea of evolution.

Over the 1750s and into the 1760s, by studying closely
related species such as the horse and the ass, Buffon did indeed
come to accept the idea that they had evolved from a common
ancestor, or (perhaps) became willing to put into print the
idea that he had actually leaned towards in 1753. He managed
to cling on to his earlier ideas about the immutability of
species and adapt them to this realisation by treating the
genera of Linnaeus (now called families) as the true species
defined by their own internal moulds. There was, for example,
an original ancestral form of the cat, which had diverged into
different forms, such as the lion, tiger and the domestic cat,
which should be regarded as varieties, not as species in their
own right. He suggested that those changes had occurred in
response to changes in diet and the environment, so that the
varieties (our species) had diverged as they migrated to
different parts of the world. But – and it is a big 'but' – there
was no suggestion that *all* life had evolved from a common
ancestor, or that humans had evolved from fish. In spite of
his earlier writings, Buffon could not bring himself to accept
the idea of a common ancestor for humans and apes. The
internal mould of each type of creature had been established
spontaneously out of 'organic particles' once the Earth cooled
to a certain point, and our mould was unique.

But these moulds did not create 'perfect' species from the beginning. Buffon pointed to the 'design' of the pig as an example of a body with room for improvement:

> [It] does not appear to have been formed upon an original, special, and perfect plan, since it is a compound of other animals; it has evidently useless parts, or rather parts of which it cannot make any use, toes all the bones of which are perfectly formed, and which, nevertheless, are of no service to it. Nature is far from subjecting herself to final causes in the formation of these creatures.

It is significant that Buffon refers to 'Nature', not to 'God'. In his 1778 book *The Epochs of Nature*, Buffon described this process occurring in stages, with new kinds of life emerging from organic particles in successive waves as the Earth cooled.* Although this idea was wrong, one aspect of it demonstrates Buffon's rejection of the idea of the Earth as a special place in the Universe with humans created by God. He believed that every planet in the Solar System (all the planets known at the time) had gone through the same process, and that the same forms of life would appear on each planet (if they had not done so already) as the planet cooled.

Hardly surprisingly, Buffon's writings provoked the ire of the religious authorities, and on more than one occasion his work was denounced from on high, by the Faculty of Theology at the Sorbonne. His reaction was always the same; he apologised, offered a written retraction of the offending passages, and promised to print the retraction in later editions of the offending book. Then he ignored the promise and continued publishing the books without the retraction. In 1785 he wrote

---

* The book was divided into seven 'epochs' to mimic the seven days of creation in Genesis.

to a friend 'I had no difficulty giving [them] all the satisfaction that [they] could desire: it was only a mockery, but men were foolish enough to be contented with it.'[14]

By the time Buffon died, in 1788, at least one evolutionary thinker in England, whose surname happened to be Darwin, was about to start publishing a series of works building up to the idea that all animal life, at least, had indeed evolved from a common ancestor, and that this process had taken hundreds of millions of years. He was able to make this bold step because in the year of Buffon's death James Hutton's *Theory of the Earth* had been published, providing hard evidence for the immense age of our planet. The work of Hutton and his successors would give the grandson of this eighteenth-century Darwin what he called 'the gift of time', a timescale long enough for evolution by natural selection to do its work, advancing by nuanced and often imperceptible degrees. This gift is so important that we need to step back from the story of biological evolution for a moment to give it the space it deserves.

CHAPTER THREE

THE GIFT
OF TIME

James Hutton's contribution to our understanding of Earth history is so important that he has been referred to as 'the father of geology'. But if that description is accurate, Robert Hooke should be known as the grandfather of geology, since it is possible to trace a clear line from his work on 'earthquakes' to Hutton's ideas. That tracing has been carried out by Ellen Tan Drake, who presents the story in her book *Restless Genius*.

Drake makes it clear that Hooke's ideas about the origin and evolution of the Earth were well known in the eighteenth century – more well known then, in fact, than they are today. One of the most intriguing threads in the story involves the German Rudolf Erich Raspe (1736 to 1794), famous now as the author of the tales of Baron Münchausen, but known in his day as what we would now call a geologist – one important enough to be elected as a Fellow of the Royal Society in 1769.

The work that got him elected was a treatise on 'the Natural History of the Terraqueous globe', which he offered as 'further corroborating the Hookian hypothesis of the Earth, on the origin of mountains and petrified bodies.' Raspe was also one of the first people to appreciate that basaltic rocks are formed from solidified lava flows. He took every opportunity to promote his (and Hooke's) ideas, not least in his translations for the Royal Society of accounts of the travels through Europe of several naturalists. Raspe embellished these with Hooke's explanation of earthquakes and volcanoes, although like his fictional creation he was not above taking much of the credit for the ideas for himself.

Another important book from the mid-eighteenth century was *The History and Philosophy of Earthquakes, from the Remotest to the Present Times, Collected from the best Writers on the Subject.* This was published anonymously in 1757, but was almost certainly the work of John Bevis (1695 to 1771), an English astronomer credited with discovering the Crab Nebula in 1771. The book was produced because of the widespread concern about earthquakes following the great earthquake of 1755, which destroyed Lisbon and killed tens of thousands of people. Just under a third of the book (106 out of 334 pages) is based on Hooke's writings, and a quote from Hooke even appears on the title page of the volume.

All this is significant because John Playfair, who promoted Hutton's work, tells us that Hutton 'carefully perused almost every book of travels from which anything was to be learned concerning the natural history of the Earth.' These must have included Raspe's translations, and Hutton can hardly have missed the Bevis book, which was published when Hutton himself was thirty-one. But for whatever reasons, in Hutton's own writings Hooke's name does not appear.

Hutton was born in Edinburgh on 3 June 1726.* His father, William, was a prominent merchant in Edinburgh, where he served as City Treasurer, and also owned two farms in Berwickshire. William died when James was still a child, and the boy was raised by his mother, who steered him towards a career as a lawyer. But after serving briefly as an apprentice to George Chalmers, he decided that the law was not for him. He was much more interested in chemistry, and at the age of eighteen he became a physician's assistant and attended lectures in medicine at the University of Edinburgh, not with any intention of becoming a doctor, but because this was the nearest he could get to studying chemistry. He also studied in Paris and Leiden, where he did receive an MD in September 1749, but decided to concentrate on developing the farms he had inherited, using the best scientific methods. He travelled around East Anglia and the Low Countries to learn the latest techniques, before settling on one of his farms and putting what he had learned into practice in the 1750s. The land he had inherited was in poor condition and covered in rocks, so with his inquisitive nature the improvement of the farm sparked his interest in geology and meteorology. But he became more than a gentleman farmer with a dilettante interest in geology, thanks to his interest in chemistry.

Together with a friend and fellow chemist, John Davie, Hutton had devised a technique for manufacturing sal ammoniac (ammonium chloride) from soot. This chemical had many important applications, including dyeing and printing (as well as in 'smelling salts'), and had previously only been available from natural sources, imported at considerable expense from the Middle East. Davie developed the technique into a

---

* The old-style calendar was still used in Britain until 1752. On our calendar, his birthday was 14 June.

practical industrial process, and both he and Hutton benefited financially as a result. While still based on the farm, in 1764 Hutton went on a tour to study the geology of the north of Scotland, accompanied by George Maxwell-Clerk, one of the antecedents of James Clerk Maxwell, the greatest physicist of the nineteenth century. But in 1768, with the proceeds of the sal ammoniac process flowing into his bank, Hutton rented out the farm and moved to Edinburgh to devote himself to science. Although this included a continuing scientific interest in farming, at the age of forty-two Hutton became in effect a full-time scientist, and a leading figure of the Scottish Enlightenment, befriending figures such as David Hume, Adam Smith and Joseph Black, and becoming one of the founders (in 1783) of the Royal Society of Edinburgh. Another of his friends was the slightly less eminent mathematician John Playfair (1748 to 1819), who would ensure that Hutton's work received due recognition.

Hutton's ideas about the Earth were developed from his personal observations of geological features on his travels around Scotland and elsewhere (as well as, of course, from his extensive reading). His most dramatic conclusion was that there was no evidence that the Earth had a finite history at all, let alone one limited to the few thousand years proposed by Bible scholars. In 1788 he concluded that there was 'no vestige of a beginning – no prospect of an end'. In other words, the Earth had always existed in more or less the same state as it is in today, and would always continue to exist in that state. This was the most extreme expression of what became known as uniformitarianism, the idea that all the features of the planet that we see today have been produced by the same natural processes that are currently at work, and which will continue to work in the same way into the indefinite future, not as a result of some great catastrophe that convulsed the Earth and produced all these features in one go.

Hutton developed his ideas over a long period of time by human standards, but he did not rush to promote them because, as John Playfair tells us, 'he was one of those who are much more delighted with the contemplation of truth, than with the praise of having discovered it'.[15] His *Theory of the Earth* was presented to the Royal Society of Edinburgh in two parts, in March and April 1785 (shortly before Hutton's fifty-ninth birthday), and expanded into book form, with additions and changes, in 1795. The book included material from his other scientific papers and pamphlets, and a good example of his thinking comes from *Concerning the System of the Earth, its Duration and Stability*, which he read to the Society on 4 July 1785:

> *The solid parts of the present land appear, in general, to have been composed of the productions of the sea, and of other materials similar to these now found upon the shores. Hence we find reason to conclude:*
>
> *1st, That the land on which we rest is not simple and original, but that it is a composition, and had been formed by the operation of second causes.*
>
> *2nd, That before the present land was made, there had subsisted a world composed of sea and land, in which were tides and currents, with such operations at the bottom of the sea as now take place. And,*
>
> *Lastly, That while the present land was forming at the bottom of the ocean, the former land maintained plants and animals; at least the sea was then inhabited by animals, in a similar manner as it is at present.*
>
> *Hence we are led to conclude, that the greater part of our land, if not the whole had been produced by operations natural to this globe; but that in order to make this land a permanent body, resisting the operations of the waters, two things had been required;*

> *1st, The consolidation of masses formed by collections*
> *of loose or incoherent materials;*
> *2ndly, The elevation of those consolidated masses from*
> *the bottom of the sea, the place where they were collected,*
> *to the stations in which they now remain above the level*
> *of the ocean.**

The image is of land being worn away by erosion, with material falling to the bottom of the sea and forming layers of sediment that are converted into rock by the weight of the material above, and raised up to form new land by geological processes, in an endless cycle. It had to be endless, a kind of perpetual motion machine, in order to fit in with Hutton's belief that God had created the world so that it would permanently be a fit home for human life. He realised that sedimentation was not the whole story when he found places where layers of granite penetrated other rocks in a way that showed that the rocks had flowed into the gaps in a molten state before setting hard. But support for his ideas came from his studies of unconformities where parallel layers of rock that had clearly been laid down horizontally had been tilted at an angle, sometimes almost to the vertical, by the forces that had lifted them up to their present position. Some of these vertical strata showed ripple marks, which clearly demonstrated that they had been laid down horizontally underwater. He explained the source of the energy required to lift and distort the solid rocks by the effect of heat flowing out from the interior of the Earth.

At the time, Hutton's model was in opposition to the more popular idea of a single great flood from which land had

---

* Compare this with Hooke: 'Parts which have been Sea are now Land, and others that have been Land are now Sea; many of the Mountains have been Vales, and the Vales Mountains.'

emerged as the waters retreated. This was known as Neptunism. Hutton's idea, dubbed Plutonism, did not initially receive as much credit as it might have done, partly because it was presented in a book more than two thousand pages long, which was written in an obscure style. But that did not stop it containing some nuggets that show the extent of Hutton's thinking, not just about the solid Earth, but about life on Earth. Try this:

> . . . *if an organised body is not in the situation and circumstances best adapted to its sustenance and propagation, then, in conceiving an indefinite variety among the individuals of that species, we must be assured, that, on the one hand, those which depart most from the best adapted constitution, will be the most liable to perish, while, on the other hand, those organised bodies, which most approach to the best constitution for the present circumstances, will be best adapted to continue, in preserving themselves and multiplying the individuals of their race.*

Crucially, he is not talking about the origin of species, but about how varieties of existing species adapt to their environment. His experience as a farmer interested in plant and animal breeding (artificial selection) had led him to this insight, but he thought that the existence of natural mechanisms of this kind was the work of a benevolent God.

Hutton died in 1797, but his ideas continued to be promoted by John Playfair, who in 1802 published his *Illustrations of the Huttonian Theory of the Earth*, partly in response to the criticisms of Hutton's work made by the Neptunists. This was a much more user-friendly book than Hutton's own writings, and reached a far wider audience. It was through Playfair that the idea of uniformitarianism first reached a wide audience, but the person who put the idea on a secure scientific footing and gave Charles Darwin the gift of time was not quite five years old when Playfair's book was published.

By that time, the foundations on which the science of geology would be built were already being laid, by the surveyor and canal builder William Smith, who had been born in the Oxfordshire village of Churchill in 1769. In the 1790s Smith was working for the Somersetshire Coal Canal Company, where his tasks included inspecting mines. He became interested in the way different layers of rock were exposed by mining and realised that not only were the strata arranged in a regular pattern, but that they could be identified by the kinds of fossils they contained. Clearly, older rocks lay beneath younger rocks, and canal excavations not only revealed the same pattern but showed how the beds were tilted at an angle and had been worn away by erosion so that older rocks lay near the surface in one place and younger rocks near the surface a little way away. In 1799, Smith made a geological map of the area around Bath, and over the next decade and a half he extended his knowledge of the geology of England and Wales while working on his own projects and various commissions as a surveyor. This culminated in the first geological map of Britain (it extended part of the way into Scotland), published in 1815. It was the first detailed geological map covering such a large area anywhere in the world, and eventually it made a major impact on science.[16] Unfortunately, Smith's business activities, including investment in a quarry to produce Bath Stone, were not so successful, and in 1819 he spent a short time in debtor's prison in London. After his release, he scraped a living as a jobbing surveyor until he was appointed Land Steward to the estate of Sir John Johnstone, in Yorkshire, in 1824. In 1831 he was the recipient of the first Wollaston Medal awarded by the Geological Society of London (their highest honour), and in

1835 he received an honorary degree from Trinity College, Dublin. He died in 1839.

Although Smith's geological map did not immediately set the scientific world on fire, even before it was published his ideas about the use of fossils in identifying strata were known among a circle of pioneering geologists, including William Buckland, who had been born in 1784 and became Reader in Mineralogy in Oxford in 1813 (he switched to being Reader in Geology in 1818). It was in this capacity that he gave lectures in the summer of 1817. These fanned the flames of the early enthusiasm of one young man in particular, Charles Lyell, who had just become interested in geology – much to the displeasure of his father, who had sent him to Oxford to study Classics and planned a career for him as a lawyer.

Lyell's father, also called Charles, had himself qualified as a lawyer, but had inherited land in Scotland and a grand house in Kinnordy at the age of twenty-six and so had no need to practise. He married in the same year that his own father died, 1796, and 'our' Charles Lyell was born in Kinnordy on 14 November 1797 – the same year that Hutton had died. But the family soon moved to a property in the New Forest, near Southampton, where the younger Charles grew up in the company of two brothers and seven sisters. After early education at a minor public school, Charles went up to Exeter College in Oxford in 1816, destined, it seemed, to follow in his father's footsteps as a lawyer and country gentleman. But that same year, he picked up a book in his father's library and became fascinated by it. The book was *An Introduction to Geology*, by Robert Bakewell, and the geology it introduced Lyell to was Hutton's uniformitarianism. Lyell went on to read Playfair's book and then to attend Buckland's lectures back in Oxford. Before this, he had had no idea that there was such a thing as the study of geology. Although he continued his study of the Classics, graduating in 1819 and receiving his MA in 1821, he became an enthusiastic amateur geologist and a Fellow of the Geological Society – which was simply a matter of being a gentleman and paying the fees. When his father took the family on an extended tour of Europe in 1818, Charles had not only been able to see the different kinds of landscape for himself, but to inspect the fossil specimens of Georges Cuvier at the Museum of Natural History in Paris. Ironically, Cuvier, who features in the next chapter, was away in England at the time. In 1821, Lyell visited the pioneering palaeontologist Gideon Mantell in Lewes, Sussex. He then returned to his legal studies in London, which he had begun the previous year, but was increasingly troubled with his eyesight and bad headaches. These were exacerbated by poring over detailed handwritten documents in poorly lit rooms, and although no formal decision to abandon the law was ever taken, and he was called to the Bar in May 1822, Lyell never practised seriously. In 1823, when he was able to visit Paris again and meet Cuvier, he was essentially a geologist.

This was emphasised by his service to the Geological Society, first as Secretary, from 1823, later as Foreign Secretary and eventually as President. In 1825, Lyell was serving as Joint Secretary to the Society, together with his exact contemporary George Scrope, who had already made a major contribution to geology and was writing a book, *Considerations on Volcanoes*, based on his expeditions to France and Italy to study extinct and active versions of the phenomenon. The two became firm friends, and Lyell soon set off on his own great geological expedition, with the intention of writing a book of his own.

Scrope had started life, on 10 March 1797, as George Thomson, but changed his name in 1821 when he married Emma Scrope (pronounced Scroop), an heiress who was the daughter of the last Earl of Wiltshire. We shall just refer to him as Scrope for consistency. George's father, John Thomson, was a wealthy trader whose firm had dealings with Russia, but little is known about the boy's early life. After attending Harrow School, Scrope went up to Pembroke College in Oxford in 1815, but quickly discovered that at the time Oxford offered inadequate opportunities to study science, and in 1816 he switched to St John's College, in Cambridge, graduating in 1821. One of his teachers in Cambridge was Adam Sedgwick, the Woodwardian Professor of Geology, who would later have a major influence on Charles Darwin. As his switch from Oxford to Cambridge highlights, Scrope was not the kind of idle gentleman undergraduate who regarded his time at university primarily as a social occasion. He was, though, a gentleman of sorts (his father claimed a connection with an aristocratic family, although his own money came from trade), and came from a family wealthy enough that even before his marriage (and while still an undergraduate) he was able to travel to Naples in the winter of 1816 to 1817. There he became intrigued by the volcano Vesuvius, and he returned on a field trip to study it in 1818. A year later he visited Etna, and in the same year that he graduated and married he travelled to study the extinct volcanoes of central France.

The volcanic origins of these mountains had been recognised in the 1750s by the Frenchman Jean-Étienne Guettard, who noted their typical cone shape, even though there has not been any volcanic activity in the region throughout recorded history. In the 1760s, his countryman Nicolas Desmarest had mapped the distribution of basaltic rocks around the Massif Central, in southern France, and shown that their patterns resembled lava flows. It was Scrope who combined these ideas with his own observations to present a coherent explanation of how that landscape had been shaped by a combination of volcanic activity and erosion.

He also witnessed first-hand a great eruption of Vesuvius in 1822. All this led to his book, *Considerations on Volcanoes*, which was published in 1825, and his election as a Fellow of the Royal Society in 1826. Scrope's book was the first to present a systematic study of volcanoes, and the first to develop a model of how volcanoes work and the part they have played in the geological history of the Earth. At the time, the book was not well received, and one of the few people who praised it was Lyell, in an essay he wrote for the *Quarterly Review* in 1827 (this was also notable as Lyell's first published essay). The problem was that Scrope's model rejected the Neptunian idea, which was the received wisdom of the time, promoted by the German geologist Abraham Werner and sometimes referred to as the Wernerian model. On that picture, the Earth had started out covered by a hot ocean containing suspended material that had gradually settled into layers to form rock, before the waters cooled and shrank to reveal the continents as they are today. Scrope saw that volcanoes were still actively contributing to the building up of land in places such as Etna, where hot material emerging from the Earth's interior formed new rock. This was the same as the kind of rock found in central France. There was no way that these rocks – basalt – could have been made by sedimentation by the processes Werner described, and no way that volcanic craters and associated geological features could have been formed by 'buckling of the Earth's crust' as the Wernerians claimed. He explained the structures he had studied in France as the product of repeated lava flows, with long intervals of calm in between outbursts of such activity, during which erosion had carved valleys into the rock. He did not estimate how long this process had taken, but clearly it required a vast stretch of time. After Scrope published his book *Geology and Extinct Volcanoes of*

Central France in 1827,* Lyell organised his own expedition to look for more evidence and settle the issue once and for all. This was to lead to Lyell's greatest work, and his gift to Darwin.

Before we pick up that thread, Scrope's own career can be briefly summarised. Although he remained active in the Geological Society – particularly active in promoting his friend Lyell's work – Scrope turned increasingly to a career in politics and as a social reformer, first as a local magistrate then, from 1833 to 1868, as a Member of Parliament. He wrote many scientific papers on geology, and was awarded the Wollaston Medal in 1867, but wrote even more pamphlets and books on political economy. After his wife died in 1867, Scrope married again, to Margaret Savage, when he was seventy and she was twenty-six. He died on 19 January 1876, a few months after Charles Lyell. His obituary in the *Proceedings of the Royal Society* said that between them Scrope and Lyell had removed geology 'from the domain of speculation to that of inductive science'. But if Scrope was the initiator of that shift, Lyell was the prime mover.

By the time he set off on his European expedition in 1828, Lyell had established a reputation as a writer, and always intended to use the expedition to gather information not just for his fellow scientists but for an accessible book on geology which he hoped would lay the Wernerian model to rest. He left England in May 1828, going to Paris to meet up with fellow geologist Roderick Murchison before the two of them went on through the Auvergne and along the southern coast of France to Italy. By September they had reached Padua.

---

* The original title was *Memoir on the Geology of Central France, including the Volcanic Formations of Auvergne, the Velay and the Vivarais*; the snappier title is from a later popularised version, but is the one usually used now to refer to the work.

Murchison then returned to England, while Lyell travelled on to Sicily, the nearest region of volcanic and earthquake activity. His field work there provided the evidence which convinced him, and eventually the whole geological community, that the features we see on Earth today have indeed been formed by the same processes visible at work today, over a very long interval of time. He already knew of work in the Massif Central by palaeontologists who had found fossil remains in what had clearly once been river sediments high above the present-day river valleys, but beneath a layer of basalt. At one site on the slopes of Mount Etna he now found the remains of seabeds more than 700 feet above sea level, sandwiched between lava flows, and unlike some of his predecessors he tried to indicate the length of time involved in producing such formations, describing:

> . . . *a very strong indication of the length of the intervals which occasionally separated the flows of distinct lava currents. A bed of [fossilised] oysters, perfectly identifiable with our common eatable species, no less than* twenty feet in thickness, *is there seen resting on a current of basaltic lava; upon the oyster bed again is superimposed a second mass of lava, together with tuff or peperino.*
>
> . . . *we cannot fail to form the most exalted conception of the antiquity of the mountain, when we consider that its base is about ninety miles in circumference; so that it would require ninety flows of lava, each a mile in breadth at their termination, to raise the present foot of the volcano as much as the average height of one lava-current.*
>
> . . . *There seems nothing in the deep sections of the Val del Bove to indicate that the lava currents of remote periods were greater in volume than those of modern times; and there are abundant proofs that the countless beds of solid rock and scoriae were accumulated, as now, in succession.*

*On the grounds, therefore, already explained, we must
infer that a mass, eight thousand or nine thousand feet in
thickness, must have required an immense series of ages
anterior to the historical periods, for its growth; yet the
whole must be regarded as the product of a modern portion
of the newer Pleistocene epoch. Such, at least, is the conclusion that we draw from the geological data already
detailed, which show that the oldest parts of the mountain,
if not of posterior date to the marine strata which are
visible around its base, were at least of coeval origin.*

The emphasis is Lyell's. As this passage highlights, Lyell realised that even a volcanic mountain such as Etna had been
built up gradually by repeated lava flows, not formed in a
single violent cataclysm as catastrophists suggested. This
extract from his *Principles of Geology* also demonstrates not
only his scrupulous attention to scientific detail, but also the
clarity of his writing, which combined to make the book a
huge success. The point he was making was clear to his readers
– Etna itself is very old by human standards, yet it sits on
rocks that are very young by geological standards, so the
Earth itself must be immensely old.

Lyell's contribution to geology is well known. It is less
well known that he also puzzled over the transmutation of
species, and discussed what might happen if 'the climate of
the highest part of the woody zone of Etna [were] transferred
to the sea-shore at the base of the mountain':

*. . . no botanist would anticipate that the olives, lemon-
tree, and prickly pear would be able to contend with the
oak and chestnut, which would begin forthwith to descend
to a lower level, or that this last would be able to stand
their ground against the pine, which would also, in the
space of a few years, begin to occupy the lower position.*

Lyell saw this as an argument *against* transmutation, intended as a refutation of the ideas of Jean-Baptiste Lamarck. Instead of the olive, lemon-tree and prickly pair mutating into new species, they would be overwhelmed by invaders already adapted to the new climate, in the same way (to use an example Lyell gave of a topical event of his day) that the native North Americans were doomed to be displaced by Europeans until one day 'these tribes will be remembered only in poetry and traditions.' Intriguingly, though, Lyell presents this example as 'a faint image of the certain doom of a species less fitted to struggle with some new condition.' Fitness, in this sense, and the struggle for survival in new conditions became, of course, the cornerstones of Charles Darwin's theory. But Lyell did accept that species went extinct and were replaced by others, although he seems to have leaned towards the idea that this was the work of a 'hands on' God:

> *Each species may have had its origin in a single pair, or individual, where an individual was sufficient, and species may have been created in succession at such times and in such places as to enable them to multiply and endure for an appointed period, and occupy an appointed place on the globe.*

Lyell did suggest that species might go extinct because of competition for resources such as food, but that new species 'took their place by virtue of a causation, which was quite beyond our comprehension'.

Lyell's book was a massive undertaking that took on board the work of geologists from across the continent of Europe, to provide a comprehensive overview of the subject. He also discussed the contents, especially of the first volume, with his friend Scrope. The title, *Principles of Geology*, was deliberately chosen as a nod to Isaac Newton's *Principia*

Mathematica (*Principles of Mathematics*), which gives you some idea of Lyell's ambitious aims. The first volume was published by John Murray, whom Lyell knew as the publisher of the *Quarterly Review*, for which journal many of his essays had been written, and appeared in July 1830. As well as the hint from the title, its ambitious aims are made clear by a subtitle: 'Being an attempt to explain the former changes in the Earth's surface, by reference to causes now in operation.' This clearly nailed Lyell's colours, if anyone was in doubt, to the mast of uniformitarianism.

Although the book was an immediate success, his eager readers had a wait for the second volume. As well as carrying out more field work, in Spain, in 1831 Lyell was appointed to the new Chair of Geology at King's College London. With all this on his mind, volume two of the *Principles* did not appear until January 1832, the year that Lyell married Mary Horner, the daughter of a geologist, who shared his interests – their honeymoon was spent on a geological tour of Switzerland and Italy. Lyell's time as a Professor at King's was a success, and he gave a series of popular lectures to which, unusually for the time, women were admitted. But he had a modest allowance from his father, and Mary brought with her a small income. Together with the income from Lyell's books and other writing, by 1833 he was financially independent, and after the publication of the third volume of his epic in that year he resigned his Chair in order to concentrate on his books and other activities. He was arguably the first professional science writer, in the sense that he had no other paid employment.

The books that Lyell concentrated on were essentially the three volumes of the *Principles*, which went through many revisions and editions. The final edition, the twelfth, appeared shortly after his death in 1875, and he had been working on it right up to the end. Lyell's other major work was a single-volume *Elements of Geology*, which was intended as a handbook for students and researchers. It first appeared in 1838 (although Lyell, being Lyell, kept revising it to keep it up to date) and became established as the first modern textbook of geology. His efforts did not go unacknowledged; he was knighted in 1848 and became a baronet (essentially, a hereditary knight) in 1864, and he has craters on both the Moon and Mars named after him.

Lyell's later life is not directly relevant to our story, except through his interaction with Charles Darwin, but it is worth digressing to show how the world was changing in the nineteenth century. In 1841 Lyell travelled by steamer to North America, where he saw the power of 'causes now in operation' at Niagara Falls and gathered new evidence of the antiquity of the Earth. He was able to travel widely by railway, and gave popular public lectures, boosting the sales of his books and his income. Lyell made three more visits to North America, travelling with an ease that would have been unimaginable half a century earlier. Indeed, he travelled with a great deal more ease than a young geologist who took a copy of the first volume of the *Principles* on a voyage round the world merely a decade earlier.

That young man, Charles Darwin, was Lyell's geological disciple, and he first made his scientific name as a result of his geological work on that voyage, aboard HMS *Beagle*, commanded by Robert FitzRoy. FitzRoy was a junior member of an aristocratic family descended from one of the acknowledged illegitimate sons of King Charles II, whom Charles had made the Duke of Grafton. As the youngest son of Lord Charles FitzRoy, Robert, who was born in 1805, could expect little (relatively speaking) in the way of inheritance, and was sent off to the Royal Naval College in Portsmouth at the age of twelve, to make his own way through a career in the Navy. He did so with such distinction that by

1828 he was serving as Flag Lieutenant to Admiral Sir Robert Otway on board the *Ganges* in South American waters. When the captain of the survey ship *Beagle* committed suicide, worn down by the pressure of work and the loneliness of command, FitzRoy was promoted to the rank of Commander and appointed as his successor. Although only a substantive Commander, he now had the courtesy title of Captain.

FitzRoy completed the surveying work that his predecessor had begun, and was back in England in the autumn of 1830. The ship was so worn out that she needed a complete refit, and FitzRoy's immediate future was uncertain. But it was soon decided to extend the survey of South America, sending FitzRoy back with the refitted *Beagle* to work down the east coast, around Tierra del Fuego, up the west coast and back home across the Pacific Ocean. The voyage would begin at the end of 1831, when FitzRoy was still only twenty-six years old. Having already experienced the loneliness of command, and uncomfortably aware of the fate of his predecessor (FitzRoy also had an uncle who had killed himself in a fit of depression), FitzRoy determined to take with him a gentleman companion who would be his intellectual and social equal, sharing his interest in the natural world, in whose company he would not be bound by the rigid requirements of naval discipline. He discussed the idea with Captain Francis Beaufort, the Hydrographer to the Admiralty, who was in overall charge of the surveying work. In the summer of 1831 Beaufort mentioned it to a friend, George Peacock, a mathematician and Fellow of Trinity College, Cambridge, who was staying in London during the university vacation. Peacock asked his colleague John Henslow, a Cambridge naturalist, if he might be interested. Henslow was thirty-five, recently married and had a new baby, so he decided that the opportunity had come about ten years too late for him, and passed it on to Leonard Jenyns, a younger Cambridge man with a growing reputation. Jenyns also turned down the proposal, having just taken up clerical duties in the church at Bottisham, a Cambridgeshire village. The deadline for the *Beagle* to sail was now uncomfortably close, and on 24 August Henslow wrote to one of Jenyn's contemporaries, Charles Darwin, in terms which left the young man little scope to refuse:

> . . . *I shall hope to see you shortly fully expecting that you will eagerly catch at the offer which is likely to be made to you of a trip to Tierra del Fuego & home by the East Indies – I have been asked by Peacock who will read & forward this to you from London to recommend him a naturalist as companion to Capt Fitzroy employed by Government to survey the S. extremity of America. I have stated that I consider you to be the best qualified person I know of who is likely to undertake such a situation – I state this not on the supposition of you being a finished Naturalist, but as amply qualified for collecting, observing & noting anything worthy to be noted in Natural History. Peacock has the appointment at his disposal & if he cannot find a man willing to take the office, the opportunity will probably be lost. Capt. F wants a man (I understand) more as a companion than a mere collector & would not take any one however good a Naturalist who was not recommended to him likewise as a gentleman.*

So who was the young man who received this letter on 29 August 1831, on his return home from a geological field trip?

Darwin had been born in Shropshire on 12 February 1809, the son of a doctor, Robert Darwin (and grandson of another doctor, Erasmus Darwin). He was the fifth of six children, with three older sisters, one younger sister, and an elder brother, Erasmus, four years his senior. His mother died in 1817, and although the two oldest girls, Marianne and Caroline, were able to supervise the running of the household (with a staff of servants), Robert became depressed and threw himself into his work as compensation for his loss. Charles had just begun attending a local day school, and in 1818 he moved on to become a boarder in Shrewsbury, where his brother was already a pupil. The two boys became very close as a result of the changes at home. In 1822 Erasmus left Shrewsbury to study medicine in Cambridge, where he was bored by lectures and became something of a party animal. When Charles was allowed to visit him in the summer of 1823, he was introduced to the life of a wealthy undergraduate, including not only drinking but the new fad, inhaling laughing gas. Back at school, Charles neglected his work, took to hunting (especially shooting birds) and generally wasted his time. So in 1825 Robert took the boy out of school and put him to work as an assistant in his medical practice. He was sufficiently impressed by an improvement in his son's attitude and an apparent interest in medicine to send him to Edinburgh to study the subject. Erasmus, who had somehow completed his three-year course in Cambridge, was off to Edinburgh at the same time for his hospital year, and Robert hoped that he would keep an eye on Charles. The two young men managed to enjoy themselves while doing the minimum of work for their formal studies, but devoted a lot of time to their real interests in science, including collecting specimens along the coast and inland.

Erasmus once again scraped through his course. But any chance of Charles becoming a doctor disappeared when he saw two operations being carried out, one on a child. There were no anaesthetics then, and the image of the screaming child remained with him for the rest of his life. He wrote in his *Autobiography*:

> *I rushed away before they were completed. Nor did I*
> *ever attend again, for hardly any inducement would have*
> *been strong enough to make me do so, this being long*
> *before the blessed days of chloroform.*
> *The two cases fairly haunted me for many a long year.*

Unable to admit how he felt to his father, after Erasmus moved on, Charles returned to Edinburgh in October 1826, ostensibly to continue his medical studies. But he actually attended classes in natural history and geology, and was strongly influenced by Robert Grant, a Scottish anatomist and expert on marine life. But in August 1827 he had to face up to the inevitable conflict with his father and confess that there was no way he could continue his medical studies and become a doctor. There was only one option left for a wastrel younger son of a respectable family who showed no inclination for the military. It was arranged that Charles would go up to Christ's College, in Cambridge, to study Classics, with the intention of becoming a country parson.

This wasn't such a bad prospect for a young man with an interest in natural history. Many country parsons had indulged in a hobby as naturalists – Gilbert White of Selborne being the prime example – and Darwin might well have followed in their footsteps. It was at Cambridge that Darwin (neglecting his official studies, of course) came under the botanical influence of John Henslow and the geological influence of Adam Sedgwick. By cramming desperately at the last minute to catch up on the work he had been neglecting, Darwin graduated respectably in 1831. Then he set off on a geological expedition around Wales, which he must have regarded as something of a last hurrah before being required to adopt the quiet life of a country vicar. But it was on his return from this expedition that he found the letter from Peacock with news of FitzRoy's proposal. His eagerness to accept the suggestion can be imagined, and although his father took some persuading before allowing Charles to set off on what seemed to him a madcap adventure, all was eventually settled and Darwin sailed with FitzRoy on board the *Beagle* on 27 December 1831, when he was still not quite twenty-three years old. He carried with him in his well-stocked library (the ship carried 245 volumes) a copy of the first volume of Lyell's *Principles of Geology* – actually a welcoming gift from FitzRoy. Henslow had advised Darwin to read the book, but 'on no account to accept the views therein.'[17] But Darwin's own observations soon persuaded him of the accuracy of Lyell's views.

The evidence was visible at the *Beagle*'s first landfall, Santiago, one of the Cape Verde islands. There, Darwin saw a band of white material, thirty feet above sea level, clearly composed of coral that had been squeezed and compressed by the weight of material above. However, coral only forms underwater. As Darwin later wrote in his *Autobiography*, 'a stream of lava formerly flowed over the bed of the sea, formed of triturated recent shells and corals which it has baked into a hard white rock.' So had the sea formerly been at least thirty feet higher than today? Or had the island risen up out of the sea? The evidence suggested to Darwin, influenced by Lyell, that the island had indeed risen; but since there were no signs of a cataclysmic event, it must mean that there had been gradual uplift over a long interval of time.

While FitzRoy and the *Beagle* were involved in months of tedious surveying work along the coast of South America, Darwin actually spent more time on land than on the ship, botanising and geologising, and sending back specimens to Henslow in Cambridge. Among the first of those specimens were the fossilised bones of a huge mammalian creature previously unknown to science. The remains of what is today known as the giant sloth caused a sensation among Henslow's scientific colleagues, and he arranged for them to be exhibited at the annual meeting of the British Association for the Advancement of Science in 1834. So the name of Charles Darwin first became widely known in scientific circles as that of a geologist and palaeontologist.

Wherever Darwin went, he found more evidence of uplift. By 1835, with the *Beagle* now exploring the western coast of South America, he was beginning to wonder whether even the mighty Andes might have been formed in this way. It was on 20 February that year that he experienced uplift first hand. He was ashore during a major earthquake, which devastated the town of Valdivia and the surrounding region. In the aftermath he saw fresh mussel beds just above the high-water level. The mussels were all dead, a yard or so above the reach even of high tide. The land had risen by that much during the earthquake. Repeated earthquakes of that kind, over a long enough time, could indeed have raised the Andes to their present height. Darwin's expeditions into the mountains confirmed this. He found fossil fishes far above sea level, petrified forests above the tree line, and jumbled geological strata showing that great forces had been at work.

But there was another side to the coin. If the Andes were rising, then, if Lyell was right, in other places the land must be sinking. Even before the *Beagle* sailed west across the Pacific, Darwin knew about the existence of coral islands, surrounded by more or less circular reefs of coral, and coral atolls where there is only the circular reef with no central island. Coral only grows in warm shallow water with plenty of sunlight. Before Darwin it was widely accepted (even by Lyell) that the reefs grew around newly formed islands, where volcanoes were rising from the sea. But Darwin realised that the opposite is true. The corals are actually fringes around islands that are gradually sinking into the sea, leaving the coral as the remnant visible at the surface. On the voyage across the Pacific, Darwin saw for himself that the young surface coral is built upon the remains of coral that has died as it sank below the waves. Although we now know that this is not the result of the whole Pacific Basin sinking downwards, Darwin's explanation of coral islands is essentially correct and also helped to make his reputation as a geologist.

Henslow had been so impressed by Darwin's work that even before the voyager returned home Henslow had had some of the letters describing his scientific findings printed up as a booklet for private circulation. In November 1835, Sedgwick had read an account of Darwin's South American discoveries to the Geological Society, and on his return to England, in October 1836, Darwin was almost immediately elected as a Fellow (he did not join the Zoological Society until 1839). On 4 January 1837 Darwin read a paper to the Geological Society describing the evidence for the gradual uplift of South America, at the rate of about one inch per century, and on 17 February, a few days after his twenty-eighth birthday, he was elected to the Society's council. The young geologist had arrived with a bang.

Darwin continued to make contributions to geology, not least with a paper on 'Volcanic Phenomena and the Elevation of Mountain Chains', presented to the Geological Society in March 1838. His detailed exposition of the evidence that the Andes had indeed been raised up by the same processes that are seen (and felt!) at work in the region today, operating over immense tracts of time, provoked a lively debate, which produced a consensus in his favour. It was after this that Lyell wrote:

I was much struck by the different tone in which my gradual causes was treated by all . . . from that which they experienced four years ago [when they were treated] with as much ridicule as was consistent with politeness in my presence.[18]

Largely on the strength of his South American work, Darwin was elected as a Fellow of the Royal Society on 24 January 1839, shortly before his thirtieth birthday. A summary of his work in South America appeared in his *Journal of Researches*,* published in May that year. The message of the book was clear, even to those who did not like his conclusions. One reviewer, intending it as a criticism, commented that, if Darwin were correct, then 'at least one million years must have elapsed' since 'the sea washed the feet of the Cordillera of the Andes'. But the critic was in a minority. Broadly speaking, the widespread scientific acceptance of gradualism and the great age of the Earth can be dated from Darwin's explanation of what he had seen in South America. For that alone he would be remembered as a significant figure in the history of science.

By the early 1840s, Darwin had a solid reputation as a geologist, and he was married and settled with his growing family in what was to be his home for the rest of his life, Down House, in a village in Kent. But he was already thinking about evolution. His interest had first been stirred by the second volume of Lyell's *Principles*, which had reached him in South America. In that volume, Lyell had given a detailed explanation of the ideas of Jean-Baptiste Lamarck, not to promote them but in order to refute them. This introduced Darwin to a version of evolutionary thinking, just at a time when he was seeing for himself the variety of the living world and the fossil evidence of species going extinct and being replaced by others. But he was cautious about going public with his developing ideas, which he knew would provoke a strong reaction, especially coming from a geologist with no reputation in the biological sciences.

---

* The full title is *Journal of Researches into the Geology and Natural History of the various Countries Visited by H.M.S.* Beagle, *under the Command of Captain FitzRoy, R. N., from 1832 to 1836.*

Before we look at how a geologist was transformed into an evolutionary biologist, though, we should bring the story of the gift of time up to date. The modern timescale of Earth's history makes even a million years seem like the blink of an eye, and provides more than enough time for evolution to do its work.

Just at the time when the evidence gathered by Lyell and Darwin was persuading geologists that the Earth really did have an immensely long history, the physicists threw a spanner in the works. The science of thermodynamics – heat and motion – developed in the nineteenth century alongside the development of steam engines. Practical experience with steam engines produced improvements in the science; improvements in the science led to improvements in the steam engines. By the middle of the nineteenth century, it was understood that although energy can be converted from one form into another (as when the heat energy in a steam engine is converted into kinetic energy of motion to make a steam engine move), such processes are not 100 per cent efficient and energy gradually leaks away into the Universe at large. This is enshrined formally in what is known as the second law of thermodynamics, and more colloquially as 'things wear out'. No store of energy is inexhaustible. And that, a few people realised in the 1840s, includes the Sun, upon which life on the surface of the Earth depends.

There were two unsung pioneers of this research whose work was not appreciated fully at the time – the German physician Julius von Mayer (1814 to 1878) and an English engineer and teacher, John Waterston (born 1811 and disappeared in mysterious circumstances in 1883). They each independently puzzled over the problem of what it is that keeps the Sun shining, and they each independently suggested that it might be 'fuelled' by a continuous supply of meteors falling onto its surface, converting gravitational energy first into the kinetic energy of the speeding meteors and then into heat energy during the impact. But their work went largely ignored, and essentially the same idea was developed by another German and another Briton, who each took it much further.

The work of William Thomson (1824 to 1907) developed the meteor impact idea to its logical conclusion, and linked it to the fate of the Earth. If the Sun itself had a finite lifetime, then, as Thomson wrote in 1852:

> *Within a finite period of time past the Earth must have been, and within a finite period of time to come the Earth must again be, unfit for the habitation of man as at present constituted, unless operations have been, or are to be performed which are impossible under the laws to which the known operations going on at the present in the material world are subject.*[19]

A year later, Thomson learned of Waterston's meteor impact idea and set about calculating how much energy would be released in such a process, and how long it could keep the Sun shining. He soon realised that meteors could not do the job, and turned his attention to the planets. He found that even if the Sun swallowed up all the planets of the Solar System one at a time, the energy released could only keep it shining for a few thousand years.

Meanwhile, Hermann von Helmholtz, who was born in Potsdam in 1821 (he died in 1894), had published his first paper on the solar energy problem in February 1854, suggesting a brilliant new idea. He proposed that the entire mass of the Sun might provide the gravitational energy to make the heat to keep it shining. If the entire mass of the Sun were spread out in the form of a cloud of rocks, bigger than the Solar System, which fell together, colliding with one another and converting gravitational energy into heat, it would produce a molten ball of fire. At the time, Helmholtz didn't calculate how much heat would be released in this way, but Thomson did, and found that it would produce as much energy as the Sun radiates in ten to twenty million years. But what was the use of releasing all that energy in one go? At first, Thomson dismissed the idea. Then, he had another thought. If, somehow, the energy could be released gradually, the Sun could shine for ten or twenty million years. Allowing a factor of ten to be on the safe side, in March 1862 Thomson wrote in *Macmillan's Magazine*:

> *It seems, therefore, on the whole most probable that the Sun has not illuminated the Earth for 100,000,000 years, and almost certain that he has not done so for 500,000,000 years. As for the future, we may say, with equal certainty, that inhabitants of the Earth cannot continue to enjoy the light and heat essential to their life, for many million years longer, unless sources now unknown to us are prepared in the great storehouse of creation.*

Thomson later developed the idea into its ultimate form. If the Sun were shrinking very slowly, it would still be releasing gravitational energy, but gradually, not all at once. A star like the Sun really can keep shining for ten or twenty million years solely by shrinking gradually and converting gravitational energy into heat. Astronomers now know that this is indeed how stars start their lives, and this timescale is known as the Kelvin-Helmholtz timescale (or in Germany as the Helmholtz-Kelvin timescale). But even the early version of the idea was enough to pose genuine problems for Darwin.

In order to demonstrate the great age of the Earth, based on uniformitarian arguments, Darwin calculated how long it must have taken for erosion to produce the landscape of the

English weald, using measurements that indicated that chalk cliffs are now being eroded at a rate of roughly one inch per century. This was only a rough and ready approximation, which gave an age that is a little on the high side compared with modern calculations, but not ludicrously so. Thomson seized on the number with something like scorn:

> *What then are we to think of such geological estimates as 300,000,000 years for the 'denudation of the Weald'? Whether is it more probable that the physical conditions of the Sun's matter differ 1,000 times more than dynamics compels us to suppose they differ from those of matter in our laboratories; or that a stormy sea, with possible channel tides of extreme violence, should encroach on a chalk cliff 1,000 times more rapidly than Mr Darwin's estimate of one inch per century?*

The difficulty troubled Darwin for the rest of his life, and led him to make some unnecessary (and unwise) revisions to his theory, which we need not go into here. The solar energy problem was actually solved, after Darwin's death, by the discovery of radioactivity, Albert Einstein's special theory of relativity, and the realisation that the Sun derives its energy from the conversion of hydrogen into helium in its heart; in fact the problem was solved precisely by 'operations' that were 'impossible under the laws to which the known operations going on at the present in the material world [were] subject,' at the time Thomson was discussing the problem. There were indeed 'sources now unknown to us . . . prepared in the great storehouse of creation'; these sources are sufficient to keep the Sun shining more or less as it does now for ten billion years, and it is now only about halfway through that lifespan. Its past history of nearly five billion years is ample for evolution to have done its work in the way Darwin described.

In parallel with the development of an understanding of the true age of the Sun and stars, following the discovery of radioactivity physicists in the twentieth century were able to determine the age of the Earth with increasing accuracy. It started with the work of Ernest Rutherford, who was born in New Zealand in 1871 but was working in Canada with English-born Frederick Soddy (1877 to 1956) when he found that radioactive atoms follow a characteristic pattern of behaviour in which half of the atoms in a sample 'decay' into something else in a certain time: the 'half-life'.* It doesn't matter how much or how little radioactive material you start with, in one half-life half of the material will decay, in the next half-life half of the remainder (one quarter of the original) will decay, and so on. The half-life is different for each kind of radioactive substance, and the material it decays into is characteristic of the substance you start with.

Radioactive uranium decays into lead, and the American Bertram Boltwood (1870 to 1927) developed a technique for determining the age of a sample of rock by measuring the proportion of lead it contains relative to different kinds (isotopes) of uranium. The technique was picked up by Arthur Holmes (1890 to 1965), who as an undergraduate at the Royal College of Science in London used it to work out the age of samples of Devonian rock from Norway as 370 million years. At the end of the first decade of the twentieth century, less than thirty years after Darwin's death, even a student could work out the ages of rocks for an undergraduate project. Holmes spent his later career refining the technique, eventually establishing that the oldest rocks (and therefore the Earth itself) had an age of 4,500 million years, neatly matching the entirely independent estimates of the age of the Sun. Along

---

* In 1907 Rutherford became Professor of Physics at the University of Manchester in England; he died in 1937.

the way, in 1944 he published a textbook, titled (in a deliberate nod to Lyell) *Principles of Physical Geology*, which became a standard text for decades. One reason for its success was its clarity – Holmes wrote to a friend that 'to be widely read in English-speaking countries think of the most stupid student you have ever had then think how you would explain the subject to him.'[20] We may not have achieved quite that level of clarity, but we trust that we have provided enough evidence that, unlike Darwin, you do not have to worry about the timescale of evolution as we return to our main theme.

PART TWO

THE MIDDLE AGES

CHAPTER FOUR

FROM DARWIN
TO DARWIN

We left our story of the evolution of the idea of evolution with the death of the Comte de Buffon and the passing of the baton to Charles Darwin's grandfather, Erasmus. Erasmus Darwin was born on 12 December 1731, the son of a retired barrister, Robert. He studied at St John's College, Cambridge, and gained an early reputation as a poet, but he had to make his own living. After further studies (including a spell in Edinburgh), he became a doctor in a village near Birmingham. Alongside his successful medical practice, Darwin developed an interest in science, and published papers on steam engines and the way in which clouds form. He married Mary Howard just after his twenty-seventh birthday, and the couple had five children. Two, Elizabeth and William, died in infancy; Charles, Erasmus and Robert all survived into adulthood (Charles only just; as a medical student in Edinburgh he cut his finger during a dissection and died from septicaemia at the age of twenty when the wound became infected). Only Robert, who was born in 1766 and became the father of 'our' Charles Darwin, married; he also became a doctor.

Robert was the youngest of the children, and he was still at home when his mother died in 1770. A seventeen-year-old girl, Mary Parker, moved into the household to look after the boy, but this wasn't all she did. She had two daughters by Erasmus, who openly acknowledged them and kept them as his own even after Mary moved out and married. Erasmus himself married a widow, Elizabeth Pole, in 1781, the year he turned fifty, and fathered another seven children with her, with six of them surviving infancy.*

You might think that all this and a thriving medical practice would leave him little time for anything else. But you would be wrong. Erasmus Darwin became a Fellow of the Royal Society in 1761, and mixed with pioneering scientists such as James Watt, Benjamin Franklin and Joseph Priestley. Hutton visited him in the summer of 1774, using Darwin's house as his base while he did some geologising in the local area. Erasmus was an eager early reader of Hutton's *Theory of the Earth* in 1788, and also one of the first people to accept the new oxygen theory of combustion; he was the principal founder of the Lunar Society, a group of scientists who met each month on the Sunday nearest the time of Full Moon, when there was light enough for them to ride home safely in the evening. He also translated Linnaeus into English. Erasmus invested wisely in new developments in canals and the iron industry, and he was a close friend of Josiah Wedgwood, founder of the eponymous pottery. Robert Darwin married

---

* An intriguing insight into Erasmus Darwin's life is provided by Charles Darwin's 'Preliminary Notice', included in Ernst Krause's assessment of the scientific work of Erasmus.

Susannah, Wedgwood's daughter, in 1796; the previous year she had inherited £25,000 when her father died, equivalent to several million pounds today. Among other things, this would mean that their son, Charles Darwin, unlike his grandfather, would never have to worry about earning a living.

Although already well respected by his peers, Erasmus gained fame outside scientific circles in his late fifties, initially through the publication in 1789 of a book called *The Loves of the Plants*. This started out as a means of popularising the work of Linnaeus in poetical form, making full use of the opportunities provided by Linnaeus for sexual allusions and innuendo. This was hot stuff in the late eighteenth century, and it reached a wide audience, including, according to Desmond King-Hele, the poets Shelley, Coleridge, Keats and Wordsworth – Coleridge certainly visited Erasmus in 1796. The success of *The Loves of the Plants* was followed, in 1792, by another poetical work, *The Economy of Vegetation*, and then by a collected edition, *The Botanic Garden*, incorporating both works. This contained 2,440 lines of verse, but underpinning the poetry there were about 80,000 words of notes, which amounted to a book about the natural world in its own right.

The stage was now set for Erasmus Darwin's greatest work, a prose book called *Zoonomia*. The first volume, some 200,000 words in length, appeared in 1794, followed by a second volume of some 300,000 words in 1796. Although most of the book is devoted to other, mostly medical, topics, in one of the forty chapters in the first volume, occupying just fifty-five pages, Darwin set out in detail his ideas on evolution, which had been skimmed over in the poetical works.

This was a dangerous time to be presenting revolutionary ideas, even in science. The French king had gone to the guillotine in 1793, and Britain was at war with France. Any threat to the established order was viewed at least with suspicion;

and often with more than suspicion. In 1790 the house of Joseph Priestley, who was an active proponent of liberal reform, as well as a pioneering chemist, was razed by rioters chanting the slogan 'Church and King Forever'; Priestley and his wife escaped, eventually to America. Evolution was certainly perceived as an anti-Church idea, and openly supporting the idea could ruin a reputation. But in 1794 Erasmus was in his sixty-third year and may have felt that it was too late to worry about his reputation, even if he felt some unease at Priestley's fate. He certainly pulled no punches, asking (clearly influenced by Hutton) whether:

> . . . *in the great length of time, since the earth began to exist, perhaps millions of ages before the commencement of the history of mankind, would it be too bold to imagine, that all warm-blooded animals have arisen from one living filament, which THE GREAT FIRST CAUSE endued with animality, with the power of acquiring new parts, attended with new propensities, directed by irritations, sensations, volitions, and associations; and thus possessing the faculty of continuing to improve by its own inherent activity, and of delivering down those improvements by generation to its posterity, world without end!*

But to what must surely have been his surprise, the chapter on evolution, simply titled 'Generation', was ignored and provoked no immediate response from reviewers or anyone else. It was buried so successfully among the pages of medical matters that as far as anyone has been able to find out, even Charles Darwin, the grandson of Erasmus, did not read it until after his own theory of evolution had been published. But a couple of years later, *Zoonomia* and its author did come under attack. Darwin was even lampooned as a revolutionary sympathiser in political cartoons. There was, at least in his own mind, a real possibility that he might be imprisoned – in 1799, Darwin's publisher, Joseph Johnson, did indeed go to jail for six months for being 'a malicious, seditious, and ill-disposed person and being greatly disaffected to our . . . sovereign Lord the King.' But Johnson really was all those things and had a track record of publishing seditious books. Darwin's *Zoonomia* was one of the mildest of them, and its author was never seriously at risk of losing his liberty.

In a pre-echo of his grandson's work, in his book Erasmus highlighted the way in which selective breeding by humans has developed new kinds of animals and plants, and noted the way in which characteristics are inherited by offspring from their parents, highlighting the example of 'a breed of cats with an additional claw on every foot'. He even noted that 'some birds have acquired harder beaks to crack nuts, as the parrot. Others have acquired beaks adapted to break the harder seeds, as sparrows. Others for the softer seeds'. But he had no idea *how* species acquired the characteristics that fitted them to their niche in the web of life. He speculated that changes were brought about in the bodies of animals and plants by their striving for something they needed, and that the characteristics they acquired in this way would be passed on to succeeding generations. In this way, a bird striving to crack hard nuts would develop a stronger beak – analogous to how a weightlifter puts on muscle. The offspring of that bird would start life with a slightly stronger beak than the one its parent was born with, and further 'striving' would result in the beaks of succeeding generations getting stronger and stronger. But one passage in particular stands out to the modern reader. Discussing the role of some species of male birds, Erasmus writes:

> *The birds which do not carry food to their young, and do not therefore marry, are armed with spurs for the*

purpose of fighting for the exclusive possession of the females, as cocks and quails. It is certain that these weapons are not provided for their defence against other adversaries, because the females of these species are without this armour. The final cause of this contest amongst the males seems to be, that the strongest and most active animal should propagate the species, which should thence become improved.

This is tantalisingly close to the idea of evolution by natural selection!

Erasmus Darwin's last book, *The Temple of Nature*, was published in 1803, and told in verse form the story of the evolution of life from the original living filament to its present diversity.* Here is a sample:

*ORGANIC LIFE beneath the shoreless waves
Was born and raised in Ocean's pearly caves
First forms minute, unseen by spheric glass,
Move on the mud, or pierce the watery mass,
These, as successive generations bloom,
New powers acquire, and larger limbs assume;
Whence countless groups of vegetation spring,
And breathing realms of fin, and feet and wing.*

and,

*Shout round the globe, how Reproduction strives
With vanquished Death – and Happiness survives;
How Life increasing peoples every clime,
And young renascent nature conquers Time.*

---

* The original title was *The Origin of Society*, but it was changed to something less provocative at the suggestion of the publisher, Joseph Johnson.

Once again, the notes were equivalent to a book in their own right, including his description of life moving out of the sea and onto the land, after the land had been raised up by volcanic activity:

> *After islands or continents were raised above the primeval ocean, great numbers of the most simple animals would attempt to seek food at the edges or shores off the new land, and might thence gradually become amphibious; as is now seen in the frog, who changes from an aquatic animal to an amphibious one . . . [organisms] situated on dry land and immersed in dry air, may gradually acquire new powers to preserve their existence; and by innumerable successive reproductions for some thousands, or perhaps millions of ages, may at length have produced many of the vegetable and animal inhabitants which people the earth.*

But Darwin had died the previous year, at the age of seventy, and was not around to promote his ideas, or to suffer any attacks for his views. The reviews were mostly hostile. Samuel Taylor Coleridge wrote to William Wordsworth that he was disgusted by the idea of 'Man's having progressed from an Ouran Outang state', which was 'contrary to all History, to all Religion, nay, to all Possibility'.[21] The *Edinburgh Review* commented that: 'If his fame be destined in anything to outlive the fluctuating fashion of the day, it is on his merit as a poet that it is likely to rest; and his reveries in science have probably no other chance of being saved from oblivion, but by having been "married to immortal verse".'

The way was left for a splendidly named Frenchman, Jean-Baptiste Pierre Antoine de Monet, Chevalier de Lamarck, to develop similar ideas in a fuller form, in which guise they became known as Lamarckism. Historians do not agree on whether Lamarck knew of Darwin's speculations or came up with the idea entirely independently (there is no evidence either way), but he certainly developed it more fully into the first comprehensive and reasonably scientific account of the way the diversity of life we see today has evolved from earlier forms. He deserves to have his name attached to the idea, but he does not deserve the mocking that those ideas have sometimes received from those who do not appreciate how significant an advance he made in the context of early nineteenth-century understanding.

In spite of his grand name (the French Chevalier is equivalent to the English Knight), Lamarck was not born with a silver spoon in his mouth. Born in Bazentin, in Picardy, on 1 August 1744, he was the eleventh child of an impoverished minor aristocrat, and he always knew that he would have to make his own way in the world. Three of his elder brothers had joined the army, the first being killed in action, and young Jean-Baptiste wanted to follow them, but his father insisted that he should enrol in a Jesuit college in Amiens. When his father died in 1760, Lamarck abandoned his studies and rode off to join the army, which was engaged in the Pomeranian War against Prussia (part of the conflict known as the Seven Years' War). The seventeen-year-old volunteer distinguished himself so much in the fighting that he was awarded a battlefield commission, but in the horseplay of the celebrations that followed, his neck was injured and he had to go to Paris for an operation, after which he spent a year recuperating. His pension was only 400 francs a year, and he stayed in Paris to study medicine for four years while working in a bank, but he gave this up to study botany, under the tuition of a leading naturalist, Bernard de Jussieu. After ten years, in 1778 he published a major three-volume work, *Flore française*, which established his reputation and led to him becoming a member of the French Academy of Sciences the following year, sponsored by the Comte de Buffon. Shortly after his thirty-fourth birthday, he had married Marie Anne Rosalie Delaporte, who would bear him six children before dying in 1792 (he would marry twice more, but all his wives predeceased him). In 1781 Lamarck was appointed as Royal Botanist, and he travelled widely over the following years to collect rare plants and other items, such as mineral samples.

In 1788 Lamarck was appointed as Keeper of the Herbarium at the Jardin du Roi, but he managed to avoid any close personal involvement with the French Revolution – it was Lamarck who wisely instigated the change of name of the botanical gardens in 1790 to Jardin des Plantes. In 1793 he became professor of what we now call invertebrate zoology at the National Museum of Natural History in Paris (it was Lamarck who gave this field that name). At the time, Lamarck believed that species did not change, but his research on molluscs led him to a different opinion. He first aired an early version of his ideas on evolution in a lecture on 11 May 1800, at the age of fifty-six, and published a book, *Hydrogéologie,* in 1802, which detailed his geological theory of the Earth. According to this model, the Earth is eternal but changing in a regular way so that it always looks much the same.* Lamarck argued that ocean currents flow from east to west, eroding material from the western borders of continents and building it up on the eastern borders of continents across the ocean, so that the continents gradually move around the globe. *Plus ça change, plus c'est le même chose.* This was uniformitarianism taken to an extreme – and it was totally wrong – but the book is remembered as one of the first to use the word biology in its modern sense, although scholars still argue about who was actually the first to use it.

---

* Even if the Earth were not eternal, its age, he said, 'utterly transcended man's capacity to calculate'.

More significantly, in the same year Lamarck published another book, *Recherches sur l'organisation des Corps Vivants*, in which his theory of evolution was developed more fully than in his lecture of 1800. This was a companion volume to *Hydrogéologie* in more than the timing of its publication. Lamarck argued that because the Earth is always changing, living things are constantly changing in order to adapt to different environments. He wrote:

> *It is the animal's habits, its mode of life, and the circum-stances in which its individual forebears found themselves that, over time, have determined the form of its body, the number and state of its organs and lastly, the faculties with which it is endowed.*

Lamarck's ideas were attacked and ridiculed by establishment figures, notably Georges Cuvier, the Professor at the Jardin des Plantes, but gained a modest following among his more junior colleagues. He continued to give his lectures, but he turned sixty in 1804 and largely avoided getting involved in public controversy, preferring to work on another book, *Philosophie zoologique*, which appeared in 1809 and set out his evolutionary ideas in detail. By then he was in poor health, with failing eyesight. In spite of this, he managed to produce an epic seven-volume work, *Histoire naturelle des animal sans vertèbres*, which appeared between 1815 and 1822. He became blind in 1818, and dependent on his surviving children, who struggled to make ends meet. When Lamarck died in 1829 they had to borrow money from the Academy of Sciences to pay for the funeral. But by then his evolutionary ideas had taken on a life of their own.

Lamarck's change of mind about the fixity of species seems to have been stimulated by his studies of simple creatures, such as molluscs. The so-called lowest forms of life had no specialised organs, and it seemed to Lamarck that they were simple enough to have been produced by spontaneous generation, as a result of the power of electricity, which was still a mysterious and poorly understood force in the 1790s and early nineteenth century. The idea that it might provide a 'life force' was taken seriously by scientists, as well as by writers such as Mary Shelley, the author of *Frankenstein* (1818). But to Lamarck it seemed necessary to have some other way of producing complex organisms, which could not be produced by spontaneous generation, and so would require some mechanism for developing complexity from simplicity. The process he invoked was more mystic than scientific. In the *Histoire naturelle*, he wrote:

> *The rapid motion of fluids will etch canals between delicate tissues. Soon their flow will begin to vary, leading to the emergence of distinct organs. The fluids themselves, now more elaborate, will become more complex, engendering a greater variety of secretions and substances composing the organs.*

The development of complexity from simplicity was described as early as the lecture of May 1800, although confusingly Lamarck put the argument upside down, saying that the invertebrates:

> *Show us still better than the others that astounding degradation in organisation, and that progressive diminution in animal faculties which must greatly interest the philosophical Naturalist. Finally they take us gradually to the ultimate stage of animalization, that is to say to the most imperfect animals, the most simply organized, those indeed which are hardly to be suspected of animality. These are, perhaps, the ones with which nature began,*

*while it formed all the others with the help of much time
and of favourable circumstances.*[22]

As the final sentence makes clear, Lamarck's argument is that
the simplest creatures emerged spontaneously and then developed into more complex forms by evolution. The reference
to 'much time' is also significant. But this is not the same as
Erasmus Darwin's view of evolution at work. Lamarck did
not accept that species went extinct, but only that forms found
in the fossil record but not living today had evolved into
forms that are still around today, and nor did he think that
all life had evolved from a single common ancestor – Erasmus
Darwin's 'filament'. Lamarck thought that new forms of life
are constantly being produced by spontaneous generation,
even today, and would develop into more complex forms as
time passes. This certainly suggests that Lamarck was unfa-
miliar with Darwin's work, if only because he did not take
the trouble to refute the latter's ideas.

Lamarck's theory was actually composed of two parts, and
the part usually referred to today as Lamarckism was the
secondary component. The primary component was what he
saw as a natural law, which encouraged, or forced, simpler
organisms to become more complex – a kind of striving for
complexity. *How* this process occurred was the secondary
consideration, involving essentially the same process that
Erasmus Darwin envisaged, with characteristics acquired by
an organism during its lifetime being passed on to succeeding
generations. But he also suggested that organs that were not
used shrank, or degraded, and eventually disappeared. In the
Philosophie zoologique, he wrote, 'lack of employment of an
organ . . . gradually impoverishes the organ and ends by
causing it to disappear entirely.' As an example of this process,
he cites the way moles have lost their sight.

Lamarck's ideas on evolution were best summed up by the four 'laws', which he presented in Volume 1 of the *Histoire*, published in 1815:

First Law: By virtue of life's own powers there is a constant tendency for the volume of all organic bodies to increase and for the dimensions of their parts to extend up to a limit determined by life itself.

Second Law: The production of new organs in animals results from newly experienced needs which persist, and from new movements which the needs give rise to and maintain.

Third Law: The development of organs and their faculties bears a constant relationship to the use of the organs in question.

Fourth Law: Everything which has been acquired . . . or changed in the organisation of an individual during its lifetime is preserved in the reproductive process and is transmitted to the next generation by those who experienced the alterations.

The fourth law is what has come to be known as Lamarckism. Perhaps the most telling point made by Lamarck, though, was the one that stuck in the throat of many of his contemporaries, and which led Charles Lyell to reject Lamarck's ideas – he specifically included humankind in the evolutionary process. Whatever the merits of the details of Lamarck's ideas, his definition of a species is hard to beat, and shows that he was indeed a profound thinker who made a real contribution to the development of evolutionary ideas:

A species is a collection of similar individuals which are perpetuated by generation in the same condition, as long as their environment has not changed sufficiently to bring about variation in their habits, their character, and their form.

Lamarck's name is often linked with that of Étienne Geoffroy Saint-Hilaire (1772 to 1844), usually referred to simply as Geoffroy. But this is largely because they both worked at the Jardin des Plantes at the same time, and not because of any real similarity in their ideas. Geoffroy thought that new forms of life could be produced by sudden leaps from one generation to the next, so that, for example, the first bird hatched from the egg of a reptile. As this example shows, he proposed that these leaps (sometimes called saltations*) happened in the embryo, and he further proposed that they are caused by changes in the environment. In a paper published in 1833, he wrote that modifications that are favourable or destructive:

. . . are inherited, and they influence the rest of the organisation of the animal because if these modifications lead to injurious effects, the animals which exhibit them perish and are replaced by others of a somewhat different form, a form changed so as to be adapted to the new environment.

Such extreme mutations, as we would now call them, have been referred to as 'hopeful monsters'. Nature is seen as flinging out a variety of saltations in the hope that one or more of them might be suited to the environment. The idea that the better adapted of these monsters survive, while others perish, is tantalisingly close to the idea of natural selection,

---

* 'Saltum' is Latin for 'jump'.

but based on the suggestion that the changes are produced in the embryo as a response to the atmosphere acting on the lungs of the parent. There are no intermediate forms, and the process is rapid (instantaneous, from one generation to the next), not gradual, even if there is then an element of selection.

But at least Geoffroy did think that evolution took place. Another of Lamarck's colleagues in Paris was violently opposed to the whole idea, even though his palaeontological work established the reality of extinctions, and is now seen as providing evidence of evolution at work.

Georges Cuvier was born in Montbéliard on 23 August 1769. He was actually christened Jean-Léopold-Nicolas-Frédéric, but an elder brother, Georges, had died earlier in 1769 at the age of four, and the new baby was always known by the old name. At the time, the town belonged to the Duchy of Württemberg, part of the Holy Roman Empire, but it became part of France in 1793. The irony is that although they ended up being bitter opponents, Lamarck and Cuvier each had half of the truth. Lamarck accepted the reality of evolution, but did not believe that there had been extinctions. Cuvier accepted the evidence of extinctions, but did not believe in evolution.

Cuvier, whose father was an officer in the Swiss Guards, became interested in natural history when he was about ten years old, and had read through the available volumes of Buffon's *Histoire Naturelle*, a copy of which was owned by an uncle, when he was twelve. He attended the local gymnasium (high school), then at the age of fifteen he moved on, thanks to the Württemberg connection, to the new Caroline Academy in Stuttgart, where he was an outstanding student. Because he had no influential contacts or private income, in 1788 he became the private tutor of Achille d'Héricy, the son of the Marquis d'Héricy, at his house in Caen. He was able to visit the botanical gardens in Caen, and the university library there. The backwater of Normandy also turned out to be a good place to be during the turmoil of the early years of the French Revolution, but in 1791 the upheaval spread there and the Marquis moved his family and his son's tutor to the relative safety of his summer house at Fiquainville. Henri Tessier, a well-known physician and expert on agriculture, had fled to Normandy under an assumed name to escape the Terror. When Tessier gave a lecture on agriculture at the town of Valmont, Cuvier recognised him and struck up an acquaintance. They soon became friends, and Tessier recognised Cuvier's ability, writing to a colleague, 'I have just found a pearl in the dunghill of Normandy.'

Under the Jacobins, Cuvier worked in an administrative role for the local commune. As the Terror subsided, Cuvier was introduced by letter to the community of naturalists in Paris, and entered into a correspondence with them. In 1795, the year the Jacobins were replaced by the Directory, the situation was calm enough for Cuvier to visit Paris with Achille, who was now nearly eighteen. The purpose of Achille's visit is not known, but Cuvier was able to make personal contact with his correspondents in Paris, which led to an invitation to work as an assistant at the Museum of Natural History, which incorporated the Jardin des Plantes, where he started work shortly before his twenty-sixth birthday.

Within a year, Cuvier had carried out his first significant piece of research, which set the tone for his career. He studied the skeletons of African and Indian elephants and compared these with each other and with the fossil remains of mammoths and a creature then known as the 'Ohio animal', but which Cuvier would later dub mastodon. In a lecture he gave in 1796, and later published, Cuvier presented the evidence that African and Indian elephants are distinct species, and that they are both different from the mammoth, implying that the mammoth has no living descendants and the species is extinct. The Ohio animal was different from all of them – another example of an extinct species. It was Cuvier's work that established once and for all the reality of extinction.

His other great contribution, closely linked to his analysis of living and fossil creatures, was to explain that all the parts of an animal's body are interdependent, and determined by its lifestyle. He spelled out this 'correlation of parts' in a paper published in 1798:

> *If an animal's teeth are such as they must be, in order for it to nourish itself with flesh, we can be sure without further examination that the whole system of its digestive organs is appropriate for this kind of food, and that its whole skeleton and locomotive organs, and even its sense organs, are arranged in such a way as to make it skilful at pursuing and catching its prey. For these reasons are the necessary conditions of existence of the animal; if things were not so, it would not be able to subsist.*[23]

This realisation was, of course, invaluable in helping Cuvier, and others, to reconstruct fossils from bits and pieces of remains. In the same paper, he continued (with some exaggeration):

> *Comparative anatomy has reached such a point of perfection that, after inspecting a single bone, one can often determine the class, and sometimes even the genus of the animal to which it belonged, above all if that bone belonged to the head or the limbs . . . up to a point – one can infer the whole from any one of [the bones].*

Cuvier's study of comparative anatomy also led him to rethink the relationships of the living world. He saw that it is not possible to represent all of life on Earth in terms of a single chain of being, or a ladder of life with 'primitive' forms at the bottom and humankind at the top. He classified animals into four groups – vertebrates, molluscs, articulates and radiates – with each of which having its own specialised anatomy. This classification is no longer used, but the fact that Cuvier divided up the animal world in such a classification scheme was a significant landmark in biological thinking, pointing the way to the analogy Charles Darwin used of a branching tree, or bush, of life.

But all this success led Cuvier's thinking up a blind alley. He saw how perfectly suited every part of an animal is to its way of life, and this led him to argue that species cannot change because any change, even to the smallest part of a creature, would be detrimental to the efficient functioning of the creature.

Cuvier became such an important figure in French science that his opposition to evolution effectively eclipsed the work of Lamarck and Geoffroy. He became a professor at the Jardin des Plantes, a foreign member of many learned societies, including the Royal Society, served in public office with distinction – both under Napoleon and after the restoration of the Bourbons – was awarded the Legion of Honour, and eventually became a Baron. By 1810, and for the rest of his life, he was arguably the most influential biologist in the world. When Cuvier spoke, the scientific world, especially in France, listened.

Cuvier was a catastrophist. As early as in the 1796 paper on elephants, he wrote:

All of these facts, consistent among themselves, and not opposed by any report, seem to me to prove the existence of a world previous to ours, destroyed by some kind of catastrophe.

As his career developed and he found more evidence of extinction of species, he decided that there must have been a series of catastrophes and, based on the limited amount of fossil evidence available to him, was convinced that new species emerged fully formed immediately after each catastrophe, remaining unchanged until they were wiped out in the next extinction. But this did not necessarily imply a new creation event after every extinction. He argued that local catastrophes might wipe out the life forms in one part of the globe, which was then repopulated by different (but not literally new) species moving in from other parts of the world. His ideas were spelled out in detail in the introduction (called a 'preliminary discourse') to a collection of his papers published in 1812; the Discourse was reprinted on its own in many languages, usually in pirated editions, and had a wide influence. Cuvier himself published an updated version in 1826 with the title *Discours sur les révolutions de la surface du globe*.

During his lifetime, Cuvier debated the idea of evolution with both Lamarck and Geoffroy. But his parting shot came from beyond the grave – indeed, from beyond the graves of both Lamarck and Cuvier. When Lamarck died in the last week of 1829, Cuvier, now sixty and a pillar of the establishment, was asked to prepare an obituary for the Academy of Sciences. This was delayed by political developments (there was rioting in Paris in 1830 in response to efforts by Charles X to turn back the tide of democracy) and by a furious debate between Cuvier and Geoffroy on the merits of 'transformation theory', which Cuvier won. His advice to young naturalists, which many took on board, was that they should confine themselves to describing the natural world, without wasting their time and effort attempting to develop theories to explain how the natural world worked. By the time he got to grips with the obituary, Cuvier was in no mood to be generous, and he savaged Lamarck's reputation. He delivered the document to the Academy at the beginning of 1832, but died in May during a cholera epidemic. The obituary was published after his death, and although titled Élogie de M. de Lamarck (Eulogy for M. de Lamarck), it summed up his ideas on evolution by saying that they:

> . . . *rested on two arbitrary suppositions; the one, that it is the seminal vapour which organises the embryo; the other, that efforts and desires may engender organs. A system established on such foundations may amuse the imagination of a poet; a metaphysician may derive from it an entirely new series of systems; but it cannot for a moment bear the examination of anyone who has dissected a viscus, or even a feather.*

He had a point, but he also threw out the fact of evolution along with Lamarck's bathwater, as an incorrect mechanism of evolution. Backed by the weight of Cuvier's reputation, this set back the development of evolutionary thinking in France just at the moment when it was gaining traction on the other side of the English Channel.

The idea was only taken on slowly, because of opposition from the conservative Establishment, including the Church; but as early as 1819 an English surgeon, William Lawrence, put into print evolutionary thoughts that significantly improved upon those of Lamarck. Lawrence was born in 1783, and lived long enough (he died in 1867) to see Charles Darwin's masterwork published. By 1819, the year in which what has come to be regarded as his own masterwork was published, Lawrence was himself a pillar of the establishment. He had been elected as a Fellow of the Royal Society in 1813, and appointed Professor of Anatomy and Surgery at the Royal College of Surgeons in 1815. His patients included Percy

Bysshe Shelley and his wife Mary, whom he also knew socially.
Lawrence was an outspoken opponent of the idea that there
is a special life force (vitalism), and his views seem to have
influenced Mary Shelley when she was writing *Frankenstein*,
which was published in 1818.

Lawrence's materialist views on the nature of life, including
human life, were published in his book *Lectures on physiology, zoology, and the natural history of man*, in 1819, when
he was in his late thirties and nearing the peak of his career.*
The fact that the book is usually referred to as *Natural history
of man*, or colloquially as 'the Lectures on Man', indicates
what it was mainly about. Lawrence was familiar with
Lamarck's work, but rejected Lamarck's mechanism for
evolution. Instead, he grasped two key features of evolution.
First, that 'offspring inherit only [their parents] connate
qualities and not any of their acquired qualities'; second, that
the differences between varieties and species (he called them
races) could only be explained by 'the occasional production
of an offspring with different characters from those of the
parents, as a native or congenital variety, and the propagation
of such varieties by generation.' What Lawrence lacked was
an explanation of the mechanism by which some varieties
were selected for survival while others failed to survive.
Nevertheless, he recognised the importance of geographical
separation in encouraging change and producing different
varieties of plants and animals, and was aware of the power
of selective breeding. In a somewhat tongue-in-cheek example
of this, he offered an explanation of why members of the
aristocracy are beautiful:

The great and noble have generally had it more in their

---

* The book included the first use of the word 'biology' in English, although
Lamarck and others had used it earlier in other languages.

> *power than others to select the beauty of nations in*
> *marriage; and thus . . . they have distinguished their*
> *order, as much by elegant proportions of person as by its*
> *prerogatives in society.*

But although this example may seem amusing to us, it highlights why the book met with such an extreme response.
Lawrence was explicitly treating humans in the same way as
the rest of the animal world. He went so far as to say that
the variations we see among people 'cannot be settled from
the Jewish Scriptures; nor from other historical records', but
must be studied using zoological techniques, and he pulled
no punches in explaining why:

> *The representations of all the animals being brought*
> *before Adam in the first instance and subsequently of*
> *their being collected in the ark . . . are zoologically impos-*
> *sible.*

He could assert this with confidence not least because
Lawrence was familiar with the work of Cuvier, Hutton and
other geologists:

> *The inferior layers, or the first order of time, contain the*
> *remains most widely different from the animals of the*
> *living creation; and as we advance to the surface there*
> *is a gradual approximation to our present species.*

And:

> *The extinct races of animals . . . those authentic memorials*
> *of beings [are] supposed, with considerable probability,*
> *to be of older date than the formation of the human race.*

As for those who believe in the literal truth of the Bible, he points out that:

> *The astronomer does not portray the heavenly motions, or lay down the laws which govern them, according to the Jewish scriptures nor does the geologist think it necessary to modify the results of experience according to the contents of the Mosaic writings. I conclude then, that the subject [origin of species] is open for discussion.*

The key features of Lawrence's work have been summarised by Cyril Darlington:

> *Mental as well as physical differences in man are inherited.*
>
> *Races of man have arisen by mutations such as may be seen in litters of kittens.*
>
> *Sexual selection has improved the beauty of advanced races and governing classes.*
>
> *The separation of races preserves their characters.*
>
> *'Selections and exclusions' are the means of change and adaptation.*
>
> *Men can be improved by selection in breeding just as domesticated cattle can be. Conversely, they can be ruined by inbreeding, a consequence which can be observed in many royal families.*
>
> *Zoological study, the treatment of man as an animal, is the only proper foundation for teaching and research in medicine, morals, or even in politics.*

All of this, but especially the inclusion of humankind as a fit subject for zoological study, was seen at the time as blasphemy. On that basis, after a strong public debate among opponents and supporters of Lawrence, in 1822 the Lord Chancellor revoked the book's copyright and Lawrence was forced to formally withdraw it from publication. As censorship this was ineffective; the book was reprinted in many pirated editions for decades. But as far as Lawrence himself was concerned, it was essentially the end of his public contribution to the debate about evolution and its significance for humankind. Faced with the prospect of an end to his career and place in society, Lawrence stuck to his medical work and soon became rehabilitated in the eyes of the Establishment. He was elected to the Council of the Royal College of Surgeons in 1828, and later became its President and 'Serjeant-Surgeon' to Queen Victoria. He even became a Baronet. In 1844 a visitor noted that the Lectures on Man:

> . . . *had interested me much some years ago, but which had rendered the author obnoxious to the clergy, because he had endeavoured to penetrate a little more deeply into the relation between the conscious and the unconscious life . . . he appears to have allowed himself to be frightened by this, and is now merely a practising surgeon, who keeps his Sunday in the old English fashion, and has let physiology and psychology alone for the present.*[24]

Curiously, two other physicians based in England had already published ideas concerning evolution, in a specifically human context, before the appearance of Lawrence's 'Lectures on Man', without incurring the opprobrium heaped on him. But these were relatively low-key presentations of the idea, in the first case in a comment so brief that a casual reader might have overlooked it. It came from James Pritchard, who had been born in Ross-on-Wye, in Herefordshire, in 1786 (he died in 1848) and studied in Edinburgh, where the theme of his doctoral thesis, presented in 1808, was the origin of the different varieties and races of people. In 1813 he published a two-volume book, *Researches into the Physical History of Man*, which was essentially a reworking and expansion of his thesis. He simply takes it for granted that human varieties have evolved from a common ancestor, writing:

> *On the whole there are many reasons which lead us to the conclusion that the primitive stock of men were prob-ably Negroes, and I know of no argument to be set on the other side.*

But if this seems a rather modest precursor of Charles Darwin, in the same year, 1813, William Wells presented what Darwin himself later described as the first recognition of the principle of natural selection.*

Wells had a colourful life prior to the presentation of that idea. He was born in Charleston, South Carolina, in 1757, to Scottish parents who had settled there in 1753. In 1775, when pressured to join the resistance to British rule, he chose to leave for Britain, where he studied medicine in Edinburgh and London. In 1779 he went to Holland, serving as a surgeon in a Scottish regiment, but fell out with his commanding officer. As a volunteer he was able to resign his commission, and promptly challenged that officer to a duel, but the man ignored the challenge. Wells then completed his medical studies at Leiden, in the Netherlands, before returning via London to Edinburgh, where he was awarded the degree of Doctor of Medicine in 1780.

The following year, Wells went back to Charleston to sort out his family's affairs there. At the time the region was still

---

* Darwin did not learn of Wells' work until after the publication of the first edition of the *Origin*; the remark appears in the 'Historical Sketch' added to later editions of the book.

controlled by the British, and Wells was able not only to tidy up his own business but also look after the interests of family friends who were now in England. When the British pulled out in 1782, Wells went with them to Florida, finally returning to England in 1784, where he settled down as a medical practitioner, becoming a Fellow of the Royal Society in 1793 and Assistant Physician at St Thomas's Hospital in 1798.

Wells' thoughts on evolution appeared in a paper read to the Royal Society in 1813, which later appeared as an Appendix to a book published five years later as *Two Essays*. But Wells had died the previous year, leaving his words to speak for themselves. The Appendix was titled 'An account of a female of the white race of mankind, part of whose skin resembles that of a negro, with some observations on the cause of the differences in colour and form between the white and negro races of man.' The passage that later met with Darwin's approval compares artificial selection (plant and animal breeding) with selection in nature, saying that what is done by animal breeders artificially ('by art'):

> . . . *seems to be done with equal efficiency, though more slowly, by nature, in the formation of varieties of mankind, fitted for the country which they inhabit. Of the accidental varieties of man, which would occur among the first few and scattered inhabitants of the middle regions of Africa, someone would be better fitted than the others to bear the diseases of the country. This race would multiply while the others would decrease . . . and as the darkest would be best fitted for the climate, this would at length become the most prevalent, if not the only race, in the particular country in which it had originated.*

This is, indeed, natural selection. But as Darwin noted, 'he applies it only to man, and to certain characteristics alone', although to be fair Wells did write 'amongst men, as well as among other animals . . .' before going into details. A much stronger statement of the general applicability of natural selection to all forms of life, and the role of the struggle for survival, appeared in 1831, shortly before Darwin set off on his voyage with Robert FitzRoy. But it appeared, once again, in an appendix, this time to a book, *On Naval Timber and Arboriculture*. It lay there unnoticed until 1860, when the author of the book pointed out its existence, the year after Darwin published his masterwork.

The author of that book, Patrick Matthew, had been born on a farm near Perth, in Scotland, in 1790. His mother, Agnes Duncan, was related to the British Admiral Adam Duncan (1731 to 1804), who won a famous victory over the Dutch at the Battle of Camperdown in 1797 and was rewarded with a peerage and land in Scotland. Matthew's family inherited an estate from Duncan, and when Robert's father died in 1807 Robert became its manager, at the age of seventeen. The land included extensive orchards, and grew grain for the whisky industry. Matthew travelled widely in Europe (in 1815 a visit to Paris had to be curtailed when Napoleon escaped from Elba) and became an expert on arboriculture. With his family connections he was especially interested in the use of timber for the construction of ships for the Royal Navy. This led him to write his book, *On Naval Timber and Arboriculture*, which appeared in the year he turned forty-one.

At the time, most naturalists still believed that species were fixed and unchanging, except for minor variations. Where evolution – or transmutation – was discussed, it was usually in the context of improving species, fitting them better to their ecological niches, as in the work of Erasmus Darwin and Lamarck. The great intellectual leap that Matthew made, pre-empting Charles Darwin and Alfred Russel Wallace, was to realise that natural selection could produce new species

– although he seems to have regarded this as so obvious that it was hardly worth making a fuss about. He even comes close to coining the term 'natural selection', referring in different places to a 'natural process of selection', a 'principle of selection', and 'selection by the law of nature'.

One of the key thrusts of his book was to criticise practices that he felt had led to a deterioration of commercially important tree species by selecting less-fit individuals, but it is his explanation of natural selection that interests us here, and his explanation is as clear as anything in Darwin or Wallace:*

> *There is a law universal in nature, tending to render every reproductive being the best possibly suited to its condition that its kind, or that organized matter, is susceptible of, which appears intended to model the physical and mental or instinctive powers, to their highest perfection, and to continue them so. This law sustains the lion in his strength, the hare in her swiftness, and the fox in his wiles. As Nature, in all her modifications of life, has a power of increase far beyond what is needed to supply the place of what falls by Time's decay, those individuals who possess not the requisite strength, swiftness, hardihood, or cunning, fall prematurely without reproducing – either a prey to their natural devourers, or sinking under disease, generally induced by want of nourishment, their place being occupied by the more perfect of their own kind, who are pressing on the means of subsistence.*

Matthew appreciated the three key ingredients of evolution by natural selection: a proliferation of individual members of

---

* This has been put in its broader context by Michael Weale, in the *Biological Journal of the Linnean Society*, Volume 115, number 4, page 1, published 19 April 2015.

species leading to competition and a 'struggle for survival', the existence of variations between individual members of a species, and the heritability of variations.

The first of these is worth picking out, as it strongly influenced not only Matthew but later evolutionary thinkers, too. The argument was put forward most forcefully by the Reverend Thomas Malthus, in a specifically human context. He was born in 1766, studied at Cambridge, and was ordained in 1788. Although he later became a Professor of History and Political Economy at Haileybury College near Hertford, it was while working as curate at Albury in Surrey that he produced the first version of his *Essay on the Principle of Population*, published in 1798. This appeared anonymously, but it was expanded into many editions published under the author's name in the nineteenth century. Malthus lived until 1834, but, like most of his contemporaries, was unaware of Matthew's work.

Malthus highlighted the fact that populations will, if given the chance, grow by geometric progression. This means that a population doubles in a certain time, then doubles again in the next interval of the same length, and so on. This applies to human populations as well as to other species. To take a simple example, if each pair of people produces four children that survive to become parents themselves, and the same thing happens in every generation, then the original couple will have sixteen grandchildren, then sixty-four great-grandchildren, and so on.* But the key is the restriction 'if every offspring survives into parenthood'. Populations remain more or less stable when the 'surplus' (Malthus' term) dies off before reproducing. Malthus specifically pointed out that at the time he was writing the human population of North America was doubling roughly

---

* But remember that each child has two parents, so the overall population is not increasing quite as dramatically as this makes it look!

once every twenty-five years, spreading out into the new lands. At that rate, the human population of the continent would reach a clearly impossible 18,000,000,000,000,000,000 in just sixteen centuries. And the same kind of argument applies to all species – dandelions, elephants, giraffes or spiders.

Malthus pointed out that populations are held in check by predators, by disease, and particularly by the amount of food available. Populations actually grow only to the limit of what can be sustained by the resources available. He wrote:

> *The natural tendency to increase is everywhere so great that it will generally be easy to account for the height at which the population is found in any country. The more difficult, as well as the more interesting, part of the inquiry is to trace the immediate causes which stop its further progress . . . What becomes of this mighty power . . . what are the kinds of restraint, and the forms of premature death, which keep the population down to the means of subsistence?*

Matthew took this on board:

> *The self-regulating adaptive disposition of organised life may, in part, be traced to the extreme fecundity of Nature, who, as before stated, has, in all the varieties of her offspring, a prolific power much beyond (in many cases a thousandfold) what is necessary to fill up the vacancies caused by senile decay. As the field of existence is limited and pre-occupied, it is only the hardier, more robust, better suited to circumstances individuals, who are able to struggle forward to maturity, these inhabiting only the situations to which they have superior adaptation and greater power of occupancy than any other kind; the weaker, less circumstance-suited, being prematurely destroyed.*

The insight that Matthew, Darwin and Wallace each hit on independently is that this process involves competition for resources among individuals, leading to the selection of the ones best suited to the environment – the fittest – to survive and reproduce while the less-fit fall by the wayside. Or as Matthew put it:

> . . . *in such immense waste of primary and youthful life, those only come forward to maturity from the strict ordeal by which Nature tests their adaptation to her standard of perfection and fitness to continue their kind by reproduction.*

The key difference between Matthew and those later writers is that, writing before Lyell's 'gift of time' was fully appreciated, he was a catastrophist and they were gradualists. He did not think that new species could arise by natural selection under the conditions we see around us on Earth today, but only after great catastrophes, matching what was known of the fossil record at the time.* And he did not think that the process could produce new complex organs:

> *Under the law of competitive selection, fins can change to feet, feet to arms, and arms to wings, and vice versa, but not this without a preordained capacity. This law guides the organs to improvement, and alters them in accommodation to circumstances should circumstances change, but cannot originate new organs. No modification of this law could originate the hollow fang of the serpent, so formed as in the forcible insertion to press upon the venom-bag at its root, and so squirt the poison into the*

---

* It is now recognised that both processes are at work. There are indeed mass extinctions followed by a proliferation of new species, but speciation also goes on in the intervals between such extinctions.

bottom of the wound; nor could it plant the rattle of warning on the tail of the most dangerous snake.

He saw evolution as proceeding in accordance with laws laid down by some designer, writing of the 'beauty and unity of design in this continual balancing of life to circumstance'. Michael Weale points out, however, that whether or not you accept the idea of the designer, the term 'law' is a much better description of natural selection than the term 'theory'. In the public mind, a law is an inescapable fact of nature, while a theory is perceived as something less certain and subject to change in the light of new evidence. By that token, natural selection is indeed a law, and Matthew realised this, writing 'there is a law universal in Nature, tending to render every reproductive being the best possibly suited to the condition that its kind . . . is susceptible of.'

Darwin, hardly surprisingly, was unaware of Matthew's publication, so he was not influenced by him. If Matthew's ideas had not been published so obscurely in a book on naval timber, Darwin might well have been encouraged by it and published his own ideas about evolution sooner. But in 1844 another book appeared that had the opposite effect on Darwin. The reception accorded to Robert Chambers' *Vestiges of the Natural History of Creation* convinced Darwin that it was not yet time for him to go public with his own evolutionary ideas.

Chambers was born in Peebles, in the Scottish border country, in 1802. His father, James, literally worked in a cottage industry, weaving cotton in a workshop on the ground floor of the house in which the family lived. Robert had an elder brother, William, and a younger brother, another James. He was educated in the basics of reading, writing and arithmetic at the local school, and went on to a high school where he was taught Classics. But to a large extent he was self-educated, reading voraciously and absorbing the contents of the *Encyclopedia Britannica* over a period of several years. By the time Robert was in high school, the family had moved to Edinburgh and William was working as a bookseller's apprentice. The move had been forced on the family by economic circumstances. First, the introduction of the power loom put cottage workers like James out of business, and he became a draper. At this time, because of the Napoleonic Wars, French prisoners on parole were housed near Peebles, and James allowed them generous credit at his store. When the prisoners were unexpectedly moved away without paying up, he went bankrupt and moved to Edinburgh to find work.

When he was sixteen, Robert left school and started to contribute to the family finances by running a bookstall on Leith Walk. He started with his father's old books, and gradually built up his stock and reputation. Meanwhile, William had bought a second-hand printing press and started his own business publishing pamphlets. In the early 1820s, the two brothers joined forces, with Robert writing and William publishing a series of cheap magazines and pamphlets that they sold for a few pennies each, and then a series of books, including a *Life of Sir Walter Scott*. In the 1830s, the brothers formally set up the publishing business W & R Chambers, while Robert also ran a bookshop in Edinburgh with his younger brother, James Junior. The two elder brothers produced a magazine, *Chambers's Edinburgh Journal*, which sold for just a penny and catered to the contemporary thirst for information about developments in science, history and the arts; its circulation quickly reached several tens of thousands and assured the financial success of their business. Robert was also a prolific contributor to the books published by W & R Chambers, which included a *Biographical Dictionary of Eminent Scotsmen*, the *Life and Works of Robert Burns*, and *Chambers's Encyclopaedia*, which was published in several volumes between 1859 and 1868. But during his lifetime Robert Chambers's name did not appear on what became his most famous work, and nor was that work published by Chambers.

Robert was fascinated by geology, and he closely followed developments from the 1830s onward; he was familiar with Lyell's work. He became a Fellow of the Royal Society of Edinburgh in 1840, and a Fellow of the Geological Society of London in 1844. He was in contact with many of the leading scientists of his day, and in 1848 he published a book, *Ancient Sea-Margins*. He later went on field trips to Scandinavia and Canada and wrote about his observations there. But by then his masterwork had already appeared.

Vestiges of the Natural History of Creation was published in 1844. The title was a deliberate nod to Hutton, who had seen 'no vestige of a beginning, – no prospect of an end'. Chambers was saying that there *had* been a beginning – that the Earth, and life on Earth, had not always existed in much the same state that we see it in today. He described a speculative model of the origin and evolution of everything, from the stars to humankind, following a progression with humankind as its culmination. This meant, among other things, that humans could not be a unique special creation but had developed – evolved – from 'lesser' animals. Chambers was well aware of the controversy that his idea was likely to provoke, and so he went to great pains to ensure his anonymity. The manuscript was copied out by his wife, so that Chambers's handwriting would not be recognised, and it was delivered to the publisher, John Churchill, in London, by a journalist friend based in Manchester, Alexander Ireland. Proofs went back by the same route, which was used from then on for all correspondence. Only three other people were in on the secret – Chambers's wife, William Chambers, and another friend, Robert Cox. Although there was much speculation and Chambers was at various times suspected of the authorship of the book, this was not formally acknowledged until after his death in 1871.

Chambers assumed that simple forms of life could be generated spontaneously, and would then evolve into more complex forms. He used his knowledge of geology to support the argument that there is a progression in the fossil record from simpler to more complex forms of life, leading up to humankind. And although he accepted that God might have set things off and established the laws by which the world (or worlds – Chambers did not think Earth was the only abode of life) operates, he specifically rejected the idea of a 'hands on' Creator tinkering with His creations:

> *Not one species of any creature which flourished before the tertiary . . . now exists; and of the mammalia which arose during that series, many forms are now altogether gone, while of others we have now only kindred species. Thus to find not only frequent additions to the previous existing forms, but frequent withdrawals of forms which had apparently become inappropriate – a constant shifting as well as advance – is a fact calculated very forcibly to arrest attention. A candid consideration of all these circumstances can scarcely fail to introduce into our minds a somewhat different idea of organic creation from what has hitherto been generally entertained.*

In other words, why should God create new species only to destroy them later? The answer must be that He set the ball rolling then left it to progress in accordance with the principles He had laid down:

> *. . . how can we suppose that the august Being who brought all these countless worlds into form by the simple establishment of a natural principle flowing from his*

mind, was to interfere personally and specifically on every occasion when a new shell-fish or reptile was to be ushered into existence on one of these worlds? Surely this idea is too ridiculous to be for a moment entertained.

Chambers did not, however, offer any mechanism for the process of evolution, other than that it was the unfolding of God's plan. Crucially, he did not see evolution as a response to changes in the environment or other external conditions. Nor did he see evolution as a continuous, gradual process; he subscribed to the idea that advances (as he saw them) were made in small jumps, or saltations. He may have been influenced in this thinking by the fact that both Robert and William Chambers had been born with an extra finger on each hand and an extra toe on each foot; the extra digits were surgically removed when the boys were infants.

In all honesty, there was nothing very new in the *Vestiges*, and anybody familiar with the work of Erasmus Darwin, Wells or Matthew would not have been surprised by it. But few people were familiar with these works, and Chambers presented his ideas as the main theme of a book, not as footnotes to an epic poem or as an appendix to a book on another topic; the anonymity also helped to create an air of mystique about them. The *Vestiges* became a sensational best-seller, and made evolution a topic of conversation in fashionable circles. It was read by Benjamin Disraeli and Abraham Lincoln, and out loud by Prince Albert to Queen Victoria. Early reviews in the popular press – and even in *The Lancet* – were favourable. But then the heavyweights of the scientific and theological establishment came down on it like a ton of bricks.

The most high-profile of many attacks on the book came from the Reverend Adam Sedgwick, one of Darwin's geological mentors, Woodwardian Professor of Geology at the University of Cambridge and now Canon of Norwich

Cathedral. Sedgwick was incensed. In a letter to Charles Lyell he called *Vestiges* a 'foul book' and wrote, 'I cannot but think the work is from a woman's pen, it is so well dressed and graceful in its externals'.[25] He was moved to write a vitriolic eighty-five-page review of the book, published in the *Edinburgh Review* in July 1845. Sedgwick expressed his concern for readers (especially 'our glorious maidens and matrons') who were being told 'that their Bible is a fable when it teaches them that they were made in the image of God – that they are the children of apes.' Such reviews were usually published anonymously, but on this occasion Sedgwick made sure he was recognised as its author.

This led in turn to attacks on Sedgwick from supporters of the contents of *Vestiges*, part of a lively debate which proved the truth of the adage that in the world of books there is no such thing as bad publicity, boosting the sales of the *Vestiges*. It appeared in numerous editions and revisions over the following decades, culminating in the twelfth edition, prepared by Alexander Ireland after Chambers's death and published in 1884 with the name of the author finally revealed.

Up to the end of the nineteenth century, *Vestiges* sold more copies than Darwin's *On the Origin of Species*. This is partly because Darwin, alarmed by the response to *Vestiges*, held off from publishing his own masterwork until 1859. He was, though, pleased to see that Sedgwick's detailed criticisms of the general idea of transmutation only raised points that he had already considered and answered in his own as yet unpublished work. He wrote to Lyell that he had been 'well pleased to find' that he had anticipated Sedgwick's objections and 'had not overlooked any of the arguments.'[26] But that did not persuade him to rush into print himself.

Although the *Vestiges* encouraged Darwin to go back into his shell, it set another naturalist on the trail that led to the theory – or law – of natural selection. That young man later wrote that it was reading Chambers's book that first persuaded him that 'transmutation' of species really did occur, and which encouraged him to plan his field work with the deliberate objective of finding evidence to back up the idea. His name was Alfred Russel Wallace, and he would be instrumental in forcing Darwin to come out of his shell and go public with his ideas.

CHAPTER FIVE

WALLACE AND DARWIN

Charles Robert Darwin and Alfred Russel Wallace each independently came up with the same Big Idea – *the* Big Idea of evolution – at about the same time. Darwin usually gets pride of place in any discussion of this idea – natural selection – but Wallace's story follows directly from the publication of the *Vestiges*, and provides the last link in the chain from the ancients to Darwin. His story logically comes first, and is, after all, every bit as interesting as that of his more famous contemporary.

Wallace was born on 8 January 1823, in a cottage just outside the town of Usk, in Monmouthshire. But his parents were English, and had arrived in this location as a result of a gradual decline in his father's fortunes. The earlier life of Thomas Vere Wallace reads like that of a minor character in a novel by Jane Austen. He had qualified as a lawyer in 1792, but, with a private income of £500 a year, he had no need to practise, and so he divided his time between London and Bath in idle pursuits. Marriage, in 1807 to Mary Anne Greenell, and the arrival of children encouraged Thomas to attempt to increase his income through a series of investments, which turned out to be unwise and had the opposite effect, prompting the move to Wales as part of a retrenchment. But when Alfred was five the family moved on again to Hertford, Mary's home town (about the same time that Charles Darwin went up to Cambridge to study for his BA). So although the Welsh like to claim Wallace as one of their own, he is best described as an Englishman born in Wales. The family suffered the kind of 'selection' common at the time. One baby girl died at the age of five months; two others at the ages of eight years and six years. Alfred was the youngest of six children who survived to adulthood, with two older brothers, William and John, and two older sisters, Elizabeth and Frances, who was known as Fanny; another brother, Herbert, was born in 1829. Soon after arriving in Hertford, Alfred met another boy, George Silk, who remained a friend for life.

At first, life in Hertford was reasonably comfortable. Although his sister Elizabeth died, at the age of twenty-two, when Alfred was nine, he says in his autobiography that he was too young to be greatly affected by it. His closest siblings, both in age and affection, were John and Fanny. With his income supplemented by taking in pupils, at first Thomas was able to provide reasonably comfortably for his own children's education. William had been apprenticed as a surveyor, and moved on to work with a large building firm in London. John also went to London, as apprentice to another builder, and Fanny went off to Lille to learn French as a prelude to becoming a teacher. Alfred was sent to Hertford Grammar School as a boarder. But then the family finances took another downward dip. Mary Wallace had inherited some money, which was put into trust for her and her children, under the administration of her sister's husband, a lawyer called Thomas Wilson. Wilson went bankrupt, and Mary's legacy was embroiled in the proceedings for years, with no income flowing from it until the legal proceedings had been untangled. Alfred was only able to complete the final year of his school education by teaching the younger boys in the school in lieu of paying full fees (about 25 guineas a year). There was no prospect of further education, and at Christmas 1836, just before he turned fourteen, he left school to make his way in the world. His parents moved to a smaller cottage in Hoddesdon, and Alfred went to London to join his brother John. It was 1837, the year Queen Victoria came to the throne. Darwin had returned to England less than a year earlier and was already thinking about evolution, although he would not read Thomas Malthus's essay until the following year.

While staying with John, Alfred spent many evenings at a so-called 'hall of science', a kind of mechanics' institute where he could read books and magazines, as well as socialise. Already well-read, this helped to crystallise his ideas about the unfairness of the social system and the puzzle of how evil and misery in the world could be reconciled with the existence of a benevolent God. But in the summer this holiday came to an end, when Alfred was apprenticed as a surveyor to his brother William. This was enjoyable work, outdoors, which gave Alfred the opportunity to learn geology and botany first hand, although the brothers barely made a living out of it. It lasted, off and on, until the beginning of 1844 (in one of the 'off' periods, Alfred was briefly apprenticed to a clockmaker, but this did not stick). During this time, Wallace read Darwin's *The Voyage of the Beagle*, as well as Lyell's *Principles of Geology*. Meanwhile, Darwin, influenced by his reading of Malthus, was, unknown to Wallace or anyone else, outlining his ideas on evolution in a series of notebooks. In one, he wrote:

> *One may say there is a force like a hundred thousand wedges trying to force every kind of adapted structure into the gaps in the economy of Nature, or rather forming gaps by thrusting out weaker ones. The final cause of all this wedgings must be to sort out proper structure & adapt it to change.*

And in October 1838:

> *Three principles will account for all*
> 1. *Grandchildren like grandparents*
> 2. *Tendency to small change especially with physical change*
> 3. *Great fertility in proportion to support of parents*

In the 1840s, the lives of both Darwin and Wallace changed. Darwin, who had married his cousin Emma in 1839, settled down with his family in the village of Down, in Kent.* For him, the days of exploration and adventure were over. For Wallace, they were just beginning. Alfred's father died in 1843, and his widowed mother had to get work as a housekeeper. Surveying work was scarce, and by the end of the year William had to let his brother go. On his twenty-first birthday, in January 1844, Alfred came into a modest inheritance of £100 from the now disentangled trust, and went to stay with his brother John in London while looking for work. He ended up taking a job as a teacher at a school in Leicester, which he was barely qualified for and ill-suited to, but it just about

---

* The village later added an 'e' to its name; so Darwin's former residence is now Down House, in Downe.

gave him enough to live on – about £40 a year. But it was here in Leicester that he read Alexander von Humboldt's *Personal Narrative* of his travels in South America, and Malthus's essay, as well as many other works in what he describes as 'a very good town library'. He also met and became friends with Henry Walter Bates. Bates was a couple of years younger than Wallace, working unenthusiastically in the family hosiery business while more enthusiastically collecting beetles and butterflies in his spare time. Wallace says that his time in Leicester was 'a turning point' because of his acquaintance with Bates, and equally because of reading Malthus.

The teaching job didn't last long. In the following winter, Alfred's brother William caught a chill while travelling in an open Third Class railway carriage at night; the chill developed into pneumonia, and he died in March 1845. Following William's funeral in Neath, Alfred stayed to wind up his business affairs and discovered that the business was faring better than he had expected. Eager to get out of the school-room, he took over and built things up on the back of the railway boom so successfully that in 1846 John joined him. They were able to rent a cottage where their mother and Herbert also came to live – Fanny was by now teaching in Macon, Georgia.

As well as surveying for proposed railway lines, the brothers designed and constructed several buildings, including a new one for the Mechanics' Institute in Neath, which is still there. There was also plenty of time for collecting, and for Alfred's continuing programme of self-education. He gave public lectures on science, and in April 1847 made his first contribution to a scientific journal when he reported the capture of a butterfly, *Carduus heterophyllus*, in the *Zoologist*. But the most significant event for his future life was that in 1845, between William's funeral and John joining him in

Neath, Alfred read the *Vestiges* for the first time. He wrote
to Bates:

> *I do not consider it as a hasty generalisation, but rather*
> *as an ingenious speculation strongly supported by some*
> *striking facts and analogies but which remains to be*
> *proved by more facts & the additional light which future*
> *researches may throw up on the subject – It at all events*
> *furnishes a subject for every observer of nature to turn*
> *his attention to; every fact he observes must make either*
> *for or against it, and it thus furnishes both an incitement*
> *to the collection of facts & an object to which to apply*
> *them when collected.*

He also read Lawrence's *Lectures on Physiology, Zoology, and
the Natural History of Man*, and wrote to Bates in December
1845, drawing attention to his argument that 'the varieties of
the Human race have not proceeded from any external cause
but have been produced by the development of certain distinctive peculiarities in some individuals which have become
propagated through an entire race.'

In 1845, Darwin's *Voyage of the Beagle* was published in
a revised edition, which included additional material, notably
on the variety of finches found on the Galapagos Islands, and
the mystery of their origin. Wallace read the new edition, and
must have noted Darwin's comment that:

> *Seeing this gradation and diversity of structure in one small,*
> *intimately related group of birds, one might really fancy*
> *that, from an original paucity of birds in this archipelago,*
> *one species had been taken and modified for different ends.*

Bates was now working as a clerk for a brewer in Burton-
on-Trent, but came on a week-long visit to Wallace in Wales,

where the two of them first hatched what Wallace refers to as a 'rather wild scheme' to make an expedition together. His 'determination to visit the tropics as a collector' was fired, Wallace tells us, by 'Darwin's *Journal* and . . . Humboldt's *Personal Narrative*.' Wild or not, Wallace clearly had the scheme in mind in the autumn of 1847, when he went to London to meet up with his sister, back from America. He spent long hours studying the collections in the British Museum, and, when he went on to Paris with Fanny, even longer hours at the Jardin des Plantes. On his return home, the idea of an expedition now seemed less wild. With Fanny at home, and developing an attachment to Thomas Sims, a photographer, their mother would be well looked after. The surveying business was less profitable, and John had decided to become a dairy farmer. Wallace was a free hand with £100 in savings, and the only problem was deciding where to go. Bates was less free; his father only reluctantly agreed to support the venture. Even so, their only hope was to finance their travels by selling the plants and creatures they collected as they went along, sending them back to an agent in London to act on their behalf. They settled on the Amazon as their destination (influenced by a book by William Edwards, *A Voyage up the River Amazon*, published in 1847), appointed Samuel Stevens as their agent, visited William Hooker, the Director of Kew Gardens, for advice, and secured a letter from him identifying them as bona fide scientific collectors. On 26 April 1848 they sailed from Liverpool. Wallace was twenty-five; Bates was twenty-three. Wallace wrote in *My Life* that even before setting out on this expedition, 'the great problem of the origin of species was already distinctly formulated in my mind . . . I firmly believed that a full and careful study of the facts of nature would ultimately lead to a solution of the mystery.'

By the time they departed on the first stage of the

adventures that would lead Wallace to discover the solution of the mystery – the law of natural selection – Darwin had essentially completed his theory, but he was not ready to publish his ideas. He had been busy since his return to England, writing up his narrative of the voyage and his geological work – including a book on coral reefs – marrying, and settling down. But he was a meticulous note-taker and he kept all his notebooks, so we have a clear record of how his ideas developed. In 1842, he wrote down, in pencil, a short 'sketch' of his thoughts about evolution. In 1844, realising that it would be many years before he got around to writing his planned book on the subject (he envisaged a three-volume epic in the pattern of Lyell's *Principles of Geology*), he decided to prepare something more formal in case he died before completing it. This was a reasonable precaution in mid-nineteenth-century England, as the fates of Elizabeth and William Wallace and their young siblings emphasise. The more formal document was an essay 230 pages long, written in ink, which the village schoolmaster had copied out in his best handwriting for Darwin. This was in July 1844, just *before* the publication of the *Vestiges*. But the reception afforded to that book helped to reinforce Darwin's determination not to formally publish his own theory until he had time to present a detailed account backed up with an unarguable weight of evidence. Contrary to some sensationalist accounts, there was nothing secret about any of this – even the village schoolmaster knew what Darwin was thinking, and Darwin discussed evolution and natural selection with close friends and colleagues. And he didn't want his ideas to be forgotten, so he left a now-famous letter for his wife with instructions on what to do with the essay if the worst happened:

> *I have just finished my sketch of my species theory. If, as I believe, my theory in time be accepted by one*

competent judge, it will be a considerable step in science, I therefore write this in case of my sudden death as my most solemn and last request . . . that you will devote 400 pounds to its publication . . . I wish this my sketch be given to some competent person, with this sum to induce him to take trouble in its improvement and enlargement.

As historian John van Wyhe has pointed out, the 'sketch' was deliberately written out with wide margins and some blank pages to allow room for 'improvement and enlargement', and Darwin clearly regarded it only as a rough draft, not yet ready to be published.

Darwin then put his work on evolution to one side and got stuck into his other tasks, finishing off his geological writings that resulted from the *Beagle* voyage and starting on a project that would keep him busy for ten more years – although he never expected it to take so long when he started. This project was a study of barnacles, which turned out to be a major contribution to natural history. If he ever flagged while carrying out this mammoth task, Darwin might have been prodded to keep at it by a comment made by the botanist Joseph Hooker* in September 1845 in a review criticising the work of a French botanist:

I am not inclined to take much for granted from anyone [who] treats the subject in his way and who does not know what it is to be a specific Naturalist himself.[27]

That comment might well have applied to the author of the *Vestiges,* as well as to the French botanist. In 1845, Darwin

---

* Joseph, who lived from 1817 to 1911, was the son of William Hooker and became one of the most influential naturalists of his time.

himself was well aware that he was known as a geologist, not as a specific naturalist – that is, someone who had studied species in detail.* He wrote to Hooker:

How painfully (to me) true is your remark that no one has hardly a right to examine the question of species who has not minutely described many.

It would be Darwin's minute description of many species of barnacle, culminating in an epic three-volume treatise published in 1854, that made him a specific naturalist and gave him the right, if he needed it, to examine the question of species. Meanwhile, that question was put to one side. Wallace sailed for the Amazon primarily in the hope of making enough money to settle down as a gentleman naturalist in England, but with the secondary hope of solving the origin of species puzzle, not knowing that Darwin had already solved it.

Wallace and Bates arrived in Brazil at the end of May 1848, and for a time worked alongside one another, Bates concentrating on collecting insects while Wallace collected specimens from plants, including trees. Like their contemporaries, they also blithely shot the wildlife as part of their collecting – and not just for collecting. Having killed a young monkey for study, rather than waste it, Wallace 'took it home, and had it cut up and fried for breakfast.'[28] The collecting went well. Their first consignment back to England included 3,635 specimens of insects (covering 1,300 different species) and twelve crates of plants for Stevens to dispose of, plus a box for William Hooker containing specimens that Wallace hoped he might purchase for Kew. An expedition by boat up the

---

* Indeed, on 30 November 1853 he would be awarded the Royal Society Medal specifically for his geological and barnacle research.

Tocantins River produced a further shipment, leading Stevens to place an advertisement in the *Annals and Magazine of Natural History* announcing the receipt of 'two beautiful Consignments . . . collected in the province of Pará, containing numbers of very rare and some new species . . . for Sale by Private Contract.'

In spite of this success, after about nine months Wallace and Bates split up and continued their collecting separately. Neither of them gave any reason for this, and they remained friends, but it seems that being in close proximity to one another all the time had become too much of a good thing. Wallace was by now also an avid collector of birds, and preparing for a much longer expedition upriver. He had also written home to suggest that his younger brother, Herbert, should join him in Brazil. Herbert was delighted. He had few prospects in Neath; his brother John was heading for the California gold fields to seek his fortune, and Fanny had married Thomas Sims and moved to Weston-super-Mare. As Herbert wrote to her, 'We are doomed to be a scatter'd family.'

He sailed for Pará on 7 June 1849, by chance on the same ship as another naturalist, Richard Spruce (1817 to 1893), who would become a lifelong friend of Alfred. For a while, Wallace, Bates and Spruce would all be collecting in the same part of the world, which had more than enough material for an army of naturalists. We do not have space here to go into many details of Wallace's adventures in the region, which were a prelude to his most important work; but we have told Spruce's story in our book *Flower Hunters*, and his experiences were very similar to those encountered by Wallace.

The brothers worked together for several months, and in the summer of 1850 they were at Barra (now Manaus), where the Rio Negro meets the Amazon. Bates and Wallace had a gentlemen's agreement that Wallace would travel up the Negro, perhaps as far as the mountains of Colombia, while

Bates worked the upper Amazon. But Herbert decided that the collecting life was not for him, and headed back downstream to Pará to get a boat home. It was an unfortunate decision; while waiting for the boat he caught a fever and died, although Alfred, far away up the Rio Negro, would not learn of his fate for months.

By the time he set off upriver in September 1850, Wallace had been in South America for two years and knew the ropes. Although he started out as a passenger on a trading vessel, he shifted to canoe and then travelled on foot (with a series of locally recruited porters), eventually reaching the mountainous region where the borders of Colombia, Venezuela and Brazil meet, not quite in territory unknown to Europeans but at the limits of von Humboldt's expedition of half a century earlier. This is the watershed that divides the drainage systems of the Negro, running down to join the Amazon and the Orinoco, flowing north and then east through Venezuela. Where possible, before reaching these remote places Wallace had sent back collections downriver for onward transmission to Stevens. These, together with the material he collected on the way back to Barra, should have produced enough money for him to settle down in England as he had planned.

Wallace was back in Barra on 15 September 1851, almost exactly a year after he had started upriver. It was then that he learned that Herbert was ill with yellow fever – in fact, he had died, at the age of twenty-two, on 8 June, but the news had not yet reached Barra. Wallace himself was ill around this time, possibly with malaria. Happily, Spruce* was based in Barra late in 1851 and early in 1852, and as Wallace recuperated they spent long hours together discussing, among other things, evolution. After recovering, Wallace made one last expedition upriver then started down for the coast with

---

* It was Spruce who broke the news of Herbert's death to Wallace.

his collections, including four large cases that should have been sent on the previous year but had been held up because they lacked the necessary paperwork.

At Pará, Wallace embarked with his cargo on the brig *Helen*, which sailed on 12 July 1852. He was almost immediately struck down with fever once again, and he spent most of his time in his cabin while he slowly recovered. Three weeks into the voyage, fire broke out in the cargo, which largely consisted of rubber; the crew and passengers had to abandon ship and take to the boats, watching the fire devour the brig. All of Wallace's collections, essentially his life's work up to that point and his prospects of a comfortable future, went up in smoke. After ten days and ten nights, burnt by the Sun and drenched by seawater, on short rations and in a desperate state, the survivors were picked up by the *Jordeson*, which turned out to be a slow, leaky old tub with barely enough food for the crew, let alone the survivors from the *Helen*.

Their troubles were far from over. In his autobiography Wallace graphically describes how, towards the end of the voyage, coming up the English Channel on 29 September, the ship nearly foundered in a violent storm. But on 1 October 1852, eighty days out from Pará, he was ashore in Deal, with the clothes he stood up in and, as he thought, nothing else in the world. Things were not, however, quite as bleak as he feared. Stevens had insured the cargo for £200. It was not enough to set Wallace up for life, but it was enough to tide him over while he planned a new expedition. And this plan would be helped by the fact that he was now, in a modest way, a known quantity to the scientific community in London and to private collectors.

The scientists knew him because, as was the usual practice in those days, Stevens had published extracts from Wallace's letters (and those of Bates) in journals such as the *Annals and*

Magazine of Natural History and the *Zoologist*. This usual practice would be very important later in the story of Wallace and Darwin. Wallace became an Associate Member of the Entomological Society, and read a couple of papers at their meetings in 1853. To have any hopes of funding another expedition, Wallace had to stay in London. By now, his brother John had returned briefly to England to get married, then returned to California, and Thomas Sims's photographic business was not doing well. Alfred arranged for his mother, Fanny and Thomas to join him in a house near Regent's Park while he planned his future. The first step was to write a small volume on *Palms of the Amazon and Rio Negro*, published at his own expense to raise his profile, and a book describing his *Travels on the Amazon and Rio Negro*, from which he was entitled to a share of the profits, although it turned out that there were no profits for nine years. He visited the collections and libraries of the British Museum (where he was on one occasion introduced to another visitor, Charles Darwin), the Linnean Society (then one of the premier learned societies) and Kew Gardens to brush up his knowledge, and in December 1852 he saw Thomas Henry Huxley give a lecture at the Zoological Society. The burning question was where he should go next, if he could get financial support.

Two factors seem to have influenced Wallace's decision to travel to the Malay Archipelago. Van Wyhe has pointed out that Stevens had handled valuable specimens from the region that had been sent back by a remarkable Viennese woman called Ida Laura Pfeiffer, who had been born in 1797 and took to travelling in 1842, after the death of her husband. After visiting the Holy Land and travelling round Europe, between 1846 and 1848 she travelled round the world, and wrote a book about her experiences. In 1851 she set off on a second trip around the globe, sending back insects from the Far East, some of which Stevens sold for her. She returned to Europe in 1854, published another book in 1855, and died in Vienna three years later. Stevens certainly would have mentioned her finds to Wallace. He also put Wallace in touch with a wealthy collector, William Wilson Saunders, who undertook to buy the bulk of Wallace's insects from the next expedition. Early in 1853, the second factor pointing Wallace east came when he made the acquaintance (we don't know exactly how) of another remarkable person, Sir James Brooke (1803 to 1868), the 'white Rajah of Sarawak'.

Brooke was a product of the British Raj – born in India but with wealthy British parents. He inherited money, bought a schooner and happened to be on hand to help the Sultan of Brunei quell an uprising. As a reward, he was given the title Rajah of Sarawak, a little corner of the large island of Borneo, and set himself up as ruler of the region, which became a free-trade port in the style of a mini Singapore, while nominally owing allegiance to the Sultan. He also served as British Consul General in Borneo and was knighted in 1847. His dynasty lasted until the Japanese invasion in World War II. Brooke has been variously cited as the inspiration for Joseph Conrad's *Lord Jim* and less plausibly Rudyard Kipling's *The Man Who Would Be King*. Brooke was in England in the spring of 1853, but shortly before he left, in April that year, he wrote to Wallace that he would be very glad to see him in Sarawak, and sent instructions to his people there to treat Wallace well if and when he did turn up. All that remained was to find a way to wangle a passage to the East.

Through the good offices of the Royal Geographical Society (RGS), which elected him as a Fellow on 27 February 1854, Wallace obtained the promise of a free passage in a Royal Navy ship, HMS *Frolic*, and was actually on board early in 1854 when the ship's orders were changed and she was sent to the Black Sea, where the Crimean War had recently broken out. Hastily offloaded, Wallace had to kick his heels until something else turned up. But the something else was worth the wait – the President of the RGS, Sir Roderick Murchison, arranged First Class travel with P&O, initially on their paddle steamer, the *Euxine*. This luxury allowed Wallace to take with him an assistant, fourteen-year-old Charles Allen, who would be accommodated rather less salubriously. They sailed on 4 March 1854, less than eighteen months after Wallace had returned to England.

Wallace's journey to Malaysia might seem slow and tortuous to modern eyes, but it was the epitome of speed and (for the most part) elegance in the 1850s. Just two decades earlier, Darwin had travelled round the world in a tiny sailing ship in conditions that would have been familiar to officers in Nelson's navy, or the navy of a century before. Now, travelling in comfort on the *Euxine*, Wallace reached Alexandria on 20 March 1854. There was time for a stopover in a good hotel, some sightseeing in Alexandria and Cairo, reached by canal, then it was onwards overland to Suez. The passengers were carried on this leg of their journey in carriages like omnibuses, each with two large wheels and pulled by four horses, making frequent stops for refreshment and to change horses (a few weeks later, a railway linking Cairo and Suez was opened; the Suez Canal opened in 1869). Waiting for them at Suez was the *Bengal*, a large liner, propelled by a single screw, which was able to carry 135 passengers in First Class comfort. She took Wallace and Allen as far as Galle, in Ceylon (modern Sri Lanka), where they transferred to another paddle steamer, the *Pottinger*, and reached Singapore on 18 April 1854, some six weeks after leaving England. All very different from Wallace's journey home from South America. His heavy baggage and equipment, travelling by the cheaper but slower route around the Cape of Good Hope, would catch up with them later, in July.

During his time on the island of Singapore, Wallace stayed for a while in Bukit Timah, outside the main town and close to the jungle, collecting insects. He was there at a time of trouble later known as the Chinese Riots, involving a rivalry between two different Chinese communities on the island, in which several hundred people were killed; but the turmoil seems to have passed him by. He then visited Malacca (where he suffered another bout of fever, treated with massive doses of quinine), but by September he was back in Singapore. Rajah Brooke was also in Singapore in September 1854, having been called before a commission investigating his rather enthusiastic anti-piracy activities (he would be exonerated). In spite of being distracted by this business, Brooke was pleased to meet up with Wallace again, and gave Wallace a letter to his nephew, John Brooke, who was in charge in Sarawak during his absence, telling him to look after Wallace until the Rajah himself could return. So Wallace and his young assistant (who was proving very able but also very lazy, according to Wallace's letters home) set off to Borneo on 17 October 1854, on board the brig *Weraff*.

Meanwhile, Darwin had finished his study of barnacles and was turning his attention to evolution again. As ever with Darwin, we know exactly what he was doing, and when. On 9 September, he noted in his journal, 'Began sorting notes for Species theory', and in his autobiography he tells us that, 'From September 1854 I devoted all my time to arranging my huge pile of notes, to observing, and experimenting, in relation to the transmutation of species.' This research involved him in pigeon breeding, as an example of artificial selection, and soliciting species from a network of contacts around the world to help him put his thoughts in order. One of those contacts would be Alfred Wallace.

We have a less complete picture of exactly what Wallace was doing and where, partly because some of his notebooks, including the one covering his first months in Sarawak, are lost, and partly because, as we shall see, he was notoriously inaccurate with his dating of the records that do survive. In all honesty, though, it must have been difficult for him to keep track of what date it was when he was away from civilisation for weeks or months on end. But you shouldn't get the idea that Wallace and Allen were alone in the trackless jungle; Wallace employed large numbers of local people as collectors, as well as the necessary servants to ensure a reasonably comfortable camp life, and ran his collecting business efficiently, sending material back to London as and when possible.

Wallace's own books about his travels are entertaining and readable, although there is more than a hint that he plays up the element of adventure to titillate his readers. Peter Raby's biography provides an excellent overview of his entire life. However, for the nitty gritty of Wallace's work in the Malay Archipelago, nothing beats van Wyhe's account, which is meticulously researched and provides the best reconstruction of the actual dates and places associated with the key events.

The first of those key events occurred in Sarawak in February 1855, when collecting was impossible because it was the rainy season, and Wallace was catching up with his reading and starting to make notes for his planned book, for which he had a working title, *The Organic Law of Change*. Part of that reading included an article by Edward Forbes, which had originally been part of his presidential address to the Geological Society on 17 February 1854 and which reached Wallace in its published form nearly a year later. Forbes proposed a variation on the theme of divine creation that we need not detail here, but which seemed so absurd to Wallace that it prompted him to write an article in riposte. The article, 'On the Law Which Has Regulated the Introduction of New Species', appeared in the *Annals and Magazine of Natural History* in August 1855, having been despatched from Sarawak by Wallace via Singapore either on 10 February on the *Weraff*, or more probably with his next consignment to Stevens, on the schooner *Dido* on 6 March.

Forbes had dismissed the idea that the fossil record showed a progression that could be seen as supporting evidence for evolution. Wallace argued that in fact the present forms of life on Earth are the products of 'a long and uninterrupted series of changes', and pointed out that the fossils show that 'Mollusca and Radiata existed before Vertebrata, and the progression from Fishes to Reptiles and Mammals, and also from the lower mammals to the higher, is indisputable'. He also emphasised the relevance of the geographical distribution of species, with closely related species in close proximity to one another in space as well as time, so that 'no species or genus occurs in two very distant localities without being also found in intermediate places', while later species in the geological record closely resemble earlier extinct species from the same part of the world. But this progression is not always along a single line; often 'two or more species have been independently formed on the basis of a common antitype [ancestor]'. So the best way to think of evolution is 'by a forked or many-branched line'. This image of a branching tree of life is one that also occurred, independently, to Darwin.

But Wallace was careful to avoid mentioning the word 'evolution' in this paper. He said that the branching tree analogy represented the 'successive creation' of species, where creation could be understood to involve natural processes, not necessarily divine intervention. Although he did not specify any mechanism by which one species succeeded another, he wrote that:

> The great law which has regulated the peopling of the Earth . . . is that every change shall be gradual; that no new creature shall be formed widely differing from

anything before existing; that in this, as in everything else in Nature, there shall be gradation and harmony.

This was a dramatic improvement on the idea of saltations – even mini-saltations. He concluded that:

Every species has come into existence coincidentally in both space and time with a pre-existing closely allied species.

This has become known as 'the Sarawak law', and although Wallace did not mention in this paper the idea of new species being descended from older species, in his own notes and in correspondence with Bates he referred to it as the law of succession of species, or simply as the succession of species. Although not spelled out, the idea was there for those with eyes to see. One of those was Charles Lyell, who was so impressed that in November 1855 he started his own notebook on species, with the name 'Wallace' at the top of page one. He also wrote to Darwin recommending the paper.

Darwin was at first less impressed, taking the term 'creation' to imply divine intervention. He wrote in the margin of his copy of the *Annals*, 'put generation for creation & I quite agree', using 'generation' as shorthand for descent from parent to offspring; in the *Origin*, however, he would write, 'I now know from correspondence that this [Wallace] attributes to generation with modification'.

It is particularly odd that Darwin missed this when he first read the paper, because Wallace included an example from Darwin's own observations in the Galapagos:

We can account for the separate islands having each their peculiar species, either on the supposition that the same original emigration peopled the whole of the islands with

the same species from which differently modified proto-
types were created, or that the islands were successively
peopled from each other, but that new species have been
created in each on the plan of the pre-existing ones.

It doesn't take much imagination to suppose that the 'plan' is passed on from one generation to the next, parent to offspring, and gets modified along the way. Wallace had presented the evidence, as van Wyhe has put it, 'in a way that was entirely consistent with an evolutionary explanation', without actually nailing his colours to the evolutionists' mast. He was testing the waters, to see what kind of reaction this might provoke. And he was establishing himself as a scientific thinker, not just a collector. All this was preparation for his planned book.

The paper did not make much impact at the time,* which rather puzzled Wallace. Perhaps he had simply been too cautious in the way he presented his ideas. But it did provoke a belated response from Darwin, in the first letter between them that has survived. In May 1857, he wrote in reply to a letter that is now lost:

By your letter & even still more by your paper in the
Annals, a year or more ago, I can plainly see that we
have thought much alike & to a certain extent have come
to similar conclusions. In regard to the Paper in Annals,
I agree to the truth of almost every word of your paper;
& I daresay that you will agree with me that it is very
rare to find oneself agreeing pretty closely with any

---

* Bates at least got the message, writing to Wallace, 'The idea is like truth itself, so simple and obvious that those who read and understand it will be struck by its simplicity; and yet it is perfectly original'. This is a curious pre-echo of Huxley's response on first learning of Darwin's theory (see page 167).

theoretical paper; for it is lamentable how each man draws his own different conclusions from the very same fact . . .

This summer will make the 20th year (!) since I opened my first note-book, on the question how & in what way do species & varieties differ from each other. – I am now preparing my work for publication, but I find the subject so very large, that though I have written many chapters, I do not suppose I shall go to press for two years.

This passage has sometimes been interpreted as a warning to Wallace that Darwin regarded the job of explaining evolution as his own private task, and the younger, less experienced, man should keep clear; but this seems implausible to us, and out of keeping with Darwin's character.* Wallace mentioned the letter when he wrote to Bates on 4 January 1858:

To persons who have not thought much on the subject I fear my paper on the 'Succession of Species' will not appear so clear as it does to you. That paper is, of course, merely the announcement of the theory, not its development. I have prepared the plan and written portions of the work embracing the whole subject . . . I have been much gratified by a letter from Darwin, in which he says that he agrees with 'almost every word' of my paper. He is now preparing his great work on 'Species and Varieties,' for which he has been collecting materials twenty years. He may save me the trouble of writing the 2nd part of my hypothesis, by proving that there is no difference in nature between the origin of species and varieties; or he may give me trouble by arriving at another conclusion; but, at all events, his facts will be given for me to work upon.

---

* In our earlier books, we have wrongly interpreted the letter as warning Wallace off; but we are now persuaded that it was nothing of the kind.

By the '2nd part of my hypothesis' Wallace clearly means the idea of what Darwin called 'generation'; even at the beginning of 1858 he had not yet hit on the idea of natural selection. There is no reason to take Darwin's letter at anything other than face value. It certainly did not deter Wallace from developing his own ideas about evolution, although in 1855 they had had to be put to one side when the rains cleared and collecting once again became his number one priority. But before he got back to work, he made a comment in one of his notebooks concerning a reference by Lyell to the 'balance of species'. This seemed to him to be 'no balance but a struggle in which one often exterminates the other'.

Wallace spent most of 1855, between March and September, collecting in Borneo, where he encountered the orangutan, the only Great Ape (apart from ourselves) found outside Africa. One of his collectors also brought in a flying frog, a creature with large webbed feet that enable it to glide (or at least fall gently) down from trees; this was previously unknown to Western science, and thereby became one of Wallace's 'discoveries', still referred to as Wallace's Flying Frog.

But in the middle of this activity, Wallace hurt his foot and was housebound for much of July and part of August, while his collectors continued their work. This gave him the opportunity for more thinking about species, and a re-reading of Lyell's *Principles of Geology*. He noted:

> *We cannot help believing the present condition of the Earth & its inhabitants to be the natural result of its immediately preceding state modified by causes which have always been & still continue in action.*

He also realised that a new species could emerge without its predecessor going extinct:

> *. . . all that the development theory requires is that some
> specimens of the lower organised group should appear
> earlier than any of the group of higher organisms.*

Even today, there are people who ask 'if people evolved from
monkeys, why are there still monkeys?'. Leaving aside the
fact that both species evolved from a common ancestor, rather
than one from the other, Wallace's comment refutes that point
on its own terms! And he asks, 'Is there more essential difference between the ass the giraffe & the zebra than between
these two varieties [greyhound and bulldog] of dogs?',
concluding that domestic varieties of dogs, hens and so on
are not regarded as distinct species because we know them
to be derived from a common stock, while 'we do not believe'
this in the case of the wild animals. That belief, he implies,
must be wrong.

Back in Sarawak, Wallace spent Christmas with Brooke
(who among other things shared Wallace's passion for chess)
and his entourage, and in January 1856 reached his thirty-
third birthday. About this time, he also received his first
communication from Darwin, in the form of a letter requesting
skins of pigeons and other birds to be sent to him. This was
not a personal communication, but one of more than two
dozen essentially identical requests sent out to collectors
around the world. Wallace's copy has not survived, but all
the copies we do have include a passage which must also have
been in his letter, and which must have caught his attention:

> *I have for many years been working on the perplexed
> subject of the origin of varieties & species, & for this
> purpose I am endeavouring to study the effects of domes-
> tication, & am collecting the skins of all the smaller
> domesticated birds & quadrupeds from all parts of the
> world.*

This puts Darwin's letter of May 1857, mentioned earlier in connection with Wallace's Sarawak paper, in perspective. It was no secret that Darwin was working on the species problem, and had been for many years; the 'revelation' of this in the 1857 letter was nothing of the kind, and would not have surprised Wallace.

Wallace also made a note of observing tiger beetles (so-called because they hunt their prey, not because of their markings) running on the shore, commenting that they were 'singularly agreeing in colour with the white sand of Sarawak'. This stuck in his mind, and would shortly become a key stimulus for his thinking on evolution.

When Wallace departed from Sarawak on 10 February 1857, he left Allen behind but took with him a servant, Ali, who would later become the competent assistant he needed. Allen would be taken under the wing of a local missionary and trained to become a teacher. Wallace's immediate destination was Singapore, where he needed to put his affairs in order and, crucially, pick up funds from Stevens for his further work. This took much longer than he had hoped, but it gave him an opportunity to write up a couple of articles (not on the species problem), to do some collecting up country, and to meet the botanical collector Thomas Lobb (1817 to 1894), working for the Veitch Nursery in England.* But while Wallace was filling in time in Singapore as best he could, things were stirring back in England. It was in the spring of 1856 that Lyell made a weekend visit to Darwin at Down House, during the course of which Darwin revealed to Lyell his idea of natural selection. Lyell, who had been one of the few people to appreciate the significance of Wallace's Sarawak paper, urged Darwin to publish at least an outline of his theory, rather than waiting to finish his big book, to establish his priority. Darwin partly

---

* Lobb also features in our book, *Flower Hunters*.

took Lyell's advice. He wouldn't publish a mere paper, but he would put off writing the big book while he prepared a short book, which he always referred to as a 'sketch' of his theory. On 14 May 1856, he noted that he 'began by Lyell's advice writing species sketch'. As the work progressed, he noted in his journal the completion of each chapter of the book.

While Darwin started on his book, Wallace was off again on his travels, via Bali, to the island of Lombok, a stopping-off point while waiting for a ship to Macassar, on Celebes (now Sulawesi). He reached Lombok on 17 June. It was there that he made an observation that alone would have assured his place in history. It had long been known that there was a distinct difference between Australian and Asian fauna and flora. One obvious example of this is the presence of marsupial mammals in some places, and placental mammals in others, even though both live under the same environmental conditions. But until Wallace came along, nobody had appreciated how sharp the dividing line between them was. In a letter to Stevens written from Lombok on 21 August 1856, Wallace discussed the geographical distribution of animals in the region:

> *The islands of Baly and Lombock, for instance, though of nearly the same size, of the same soil, aspect, elevation and climate, and within sight of each other, yet differ considerably in their productions, and, in fact, belong to two quite distinct ecological provinces, of which they form the extreme limits.*

An extract from Wallace's letter* was published by Stevens in the *Zoologist* in January 1857, following the usual practice

---

* The letter, sent with a consignment of specimens, also included the first mention of his response to Darwin's request for birds in Wallace's correspondence: 'The domestic duck . . . is for Mr. Darwin'.

with interesting news from far-flung places. On his further travels, Wallace continued to study the demarcation between these two 'ecological provinces' with his surveyor's eye, and in January 1858 in a letter to Bates he referred to there being a 'boundary line' between them. It was, however, only after he was back in England that he published a paper, in 1863, which included a map of the region with the boundary line marked in red. This became known as the Wallace Line, although the exact position of the line has since been adjusted in the light of later studies.

Wallace thought that the existence of the two provinces was a result of the breakup of two former continents that had sunk beneath the sea. We now know that the uniquely Australian species evolved when there was a much larger separation between Australia and Asia, and were carried to their present location over millions of years by the slow processes of plate tectonics – by horizontal, not vertical, movement. Wallace's line is important both for our understanding of the evolution of life and for our understanding of the geological evolution of the planet.

In September, Wallace moved on to the Dutch settlement of Macassar. This was a thoroughly civilised town, but not immune from fever (almost certainly malaria), which struck down both Wallace and Ali during their stay. In October, Wallace wrote a letter to Darwin, which has been lost. Darwin's reply, dated 1 May 1857, is the letter we mentioned earlier. Judging from this, there was nothing significant in Wallace's letter from our point of view, but it shows that the two had now developed a direct personal correspondence.

On 18 December Wallace left Macassar, heading east on a thousand-mile journey to the Aru Islands, where he would stay until July 1857. His main objective was to find and collect birds of paradise, both for their interest and because their skins would fetch large prices back in England. This objective was amply fulfilled; in a letter to Stevens he wrote, 'I believe I am the only Englishman who has ever shot and skinned (and ate) birds of Paradise'. But the beauty of the birds posed a puzzle, which he addressed in his book *The Malay Archipelago*:

> *It seems sad that on the one hand such exquisite creatures should live out their lives and exhibit their charms only in these wild inhospitable regions . . . This consideration must surely tell us that all living things were not made for man.*

Wallace returned to Macassar in July, and sent Stevens his collections from Aru, which would eventually bring in the best part of a thousand pounds. Then he settled down to making more notes for his planned book. A key passage reads, 'All varieties we know are produced at *birth* the offspring differing from the parent. This offspring propagates its kind.' He also wrote up his observations on the species found in the Aru Islands. But most importantly for our understanding of the development of his ideas on evolution, he wrote a 'Note on the theory of permanent and geographical varieties', which would appear in the *Zoologist* in January 1858. He asks the question 'What is a *species*?', then says:

> *A species differs from a variety in degree only, not in nature . . . the line that separates them [is] so fine that it will be exceedingly difficult to prove its existence.*

The correspondence with Darwin was also proving fruitful. Between September and November 1857, Wallace was based a little way inland, up the Maros River. On 27 September, he wrote a letter to Darwin from which a tantalising fragment survives because Darwin cut it out to keep a note about jaguars written on the other side of the paper. After thanking Darwin for his encouraging letter of May that year, he confesses that he had been disappointed by the lack of response to his Sarawak paper, and says that this was:

> . . . *of course but preliminary to an attempt at a detailed proof of [my theory], the plan of which I have arranged, & in part written, but which of course requires much research in English libraries & collections.*

If Darwin did have any concerns about Wallace beating him into print (which seems unlikely), he would have been reassured by this letter, which suggested that, like Darwin himself, Wallace was in no hurry and that he would not be completing his own book until he returned home.

Back at Macassar, Wallace came across some more tiger beetles. These lived on shiny brown mud and were so nearly the same colour as the mud that he could only detect them by the shadows they cast. In Sarawak, white tiger beetles lived on white sand; in Macassar, brown tiger beetles lived on brown mud, each perfectly matching their background. The discovery stuck in his mind, although as yet he had no explanation for it.

On 19 November, Wallace left Macassar and made his way in leisurely stages to the small island of Ternate, 48 miles north of the Equator, where the pieces of the species puzzle would at last fall into place. He arrived there on 8 January 1858. After settling into a rented house, which would be his main base for the next three years, he prepared for an expedition to the nearby island of Gilolo (now known as Halmahera), but before he could set out he was struck by another bout of fever. The chronology of what happened next is confused because the dates given by Wallace in his later writings do not always agree with the dates given in the journals at the time, and some of those dates are not only inconsistent but plainly wrong, as when he writes '20 January' when he means '20 February'. But the actual course of events has been painstakingly reconstructed by van Wyhe, shedding light on the most important few weeks in Wallace's life.

Wallace's own account, from *My Life*, stresses the importance of Malthus's essay on his flash of insight. In early February 1858 he was suffering from intermittent bouts of fever, and had to postpone his visit to Gilolo while he recovered:

> *One day something brought to my recollection Malthus's 'Principles of Population', which I had read about twelve years before. I thought of his clear exposition of 'the positive checks to increase' – disease, accidents, war, and famine – which keep down the population of savage races to so much lower an average than that of more civilized peoples. It then occurred to me that these causes or their equivalents are continually acting in the case of animals also; and as animals usually breed much more rapidly than does mankind, the destruction each year from these causes must be enormous . . . it occurred to me to ask the question, Why do some die and some live? And the answer was clearly, that on the whole the best fitted live . . . The more I thought over it the more I became convinced that I had at length found the long-sought-for law of nature that solved the problem of the origin of species.*

After making notes 'the same evening', Wallace then spent two days carefully writing out his theory, in the form of an essay with the title 'On the Tendency of Varieties to Depart Indefinitely from the Original Type'; he later recalled that this was in order to send it to Darwin by the next available post – not because he thought Darwin would be the ideal person to see it, but because he wanted to ask Darwin to pass it on to Lyell, with whom Wallace was not in direct communication, but who he thought would be the best person to offer an opinion. This explanation is open to some doubt, and the essay may originally have been written simply because he wanted to get it down while the ideas were fresh in his mind, ready to be elaborated on in his planned book. The essay was dated 'Ternate, February, 1858', so we don't even know the exact date on which it was completed. A more intriguing question, though, is: what was the 'something' that started Wallace thinking about Malthus's essay?

The most likely trigger was his observations of tiger beetles. On 2 March, only a couple of weeks after completing his essay, Wallace wrote to Bates that two sets of beetles he had studied:

> . . . *are sea beach insects . . . the former singularly agreeing in colour with the white sand of Sarawak, the latter with the dark volcanic sand of its habitat. Others prefer river banks . . . another . . . was found in the soft shiny mud of salt creeks, with which its colour so exactly agrees that it was perfectly invisible except for its shadow. Such facts as these puzzled me for a long time, but I have lately worked out a theory which accounts for them naturally.*

The final sentence clearly links the realisation of the idea of natural selection with the appearance of the tiger beetles. And in the essay itself, he writes:

> . . . *the peculiar colours of many animals, especially insects, so closely resembling the soil or the leaves or the trunks on which they habitually reside, are explained on the same principle; for though in the course of ages varieties*

> *of many tints may have occurred, yet those races having colours best adapted to concealment from their enemies would inevitably survive the longest.*

This is evolution by natural selection in a nutshell. It is not chance that decides which individuals in a generation die and which survive; those that live and reproduce in their turn must be the ones best suited to the prevailing conditions. If beetles of many hues were living on a surface of a particular colour, predators would easily pick out and eat the ones that looked least like their background. The survivors would reproduce and produce offspring which by and large would resemble the background better than the previous generation, but once again the less well-camouflaged would be picked off first. After many generations, the result would be a colour that 'so exactly agrees [with the background] that it was perfectly invisible except for its shadow'.

There was, as we shall see, much more to the essay. But why should we doubt that it was originally intended for Darwin and Lyell? And when was it sent to them?

Wallace returned from Gilolo on 1 March, and eight days later mail arrived from England. This included a letter from Darwin dated 22 December 1857, written in response to Wallace's letter of 27 September, reassuring Wallace that his Sarawak paper had been noticed by 'good men', including Lyell, and commenting 'though agreeing with you on your conclusion in that paper, I believe I go much further than you, but it is too long a subject to enter on [here]'. This was the first Wallace knew that Lyell had shown an interest in his work. It was this letter that prompted him to reply to Darwin, enclosing his essay and, as he explained in *My Life*, asking him 'if he thought it sufficiently important, to show it to Sir Charles Lyell, who had thought so highly of my former paper'.

Thanks to van Wyhe's reconstruction, we can follow the

course of the Ternate essay as our focus shifts from Wallace in the Malay Archipelago to Darwin in a village in Kent. Wallace was off on his travels again on 25 March, this time to New Guinea, but he left behind the package for Darwin to be sent on the next mail steamer, which departed from Ternate on 5 April. A succession of mail boats carried it via Surabaya, Batavia and Singapore to Galle, where it arrived on 10 May. From Galle the sea route took it to Suez, arriving on 3 June, then overland to Alexandria. The steamer *Colombo* left Alexandria on 5 June and reached Southampton on 16 June; the mail got to London on 17 June, and Darwin noted that he received the package containing Wallace's letter and essay at Down House on 18 June 1858, some 75 days after it left Ternate. Less than a fortnight later, on 1 July, some five months after it was written, the essay would be presented to the scientific community at large. Those were a busy two weeks for Darwin and his friends.

CHAPTER SIX

DARWIN AND WALLACE

When Wallace's Ternate essay arrived at Down House, Darwin was well on the way to completing his 'sketch' of his species theory – the theory of evolution by natural selection. Just how far on the way, and how much Wallace's communication must have come as a shock, we can see by summarising the extent of his progress up to that point.

By the time he returned from the *Beagle* voyage, Darwin was convinced of the fact of evolution. The problem that confronted him, and other evolutionists, was discovering the mechanism by which evolution worked. His first note-book on *The Transformation of Species* was begun in 1837, and a year later he read Malthus's essay. It was this that led Darwin to the realisation that the pressure that drives evolution is the struggle for survival involving competition among members of the *same* species, not the competition between species. A lion is not competing with the prey it feeds upon, but with other lions for the ability to catch prey; the prey is not competing with the lion, but with other members of its own species to escape from the lion. This is the truth behind the old joke about two hunters being chased by a grizzly bear – neither of them can run faster than the bear, but the one who runs faster than the other hunter will survive. Darwin sketched out these ideas in a document that historians have dated to 1839, and then in the pencil sketch of 1842, which developed into the more formal document of 1844.

But that wasn't all; we also have a curious insight into Darwin's thinking around that time about his big idea and his desire to make sure that it would not be lost if anything happened to him, as well as his natural wish to ensure that his priority would be recognised. We owe this to the detective work of Howard Gruber, an American psychologist who was fascinated by the nature of creative thinking and who made a special study of how Darwin worked.[29] In the second, 1845, edition of the book that we know as *The Voyage of the Beagle*, but which was actually titled *Journal of Researches*, Darwin added a lot of new material, as we have mentioned, scattered here and there through the pages. The new edition was published three years after Darwin completed his pencil 'sketch' and just a year after the more formal essay was copied out by the village schoolmaster. The revision was also completed just before Darwin embarked on his epic study of barnacles. All this was obviously a clearing of the decks prior to tackling that great work, and thanks to Gruber we can see that even the revision of *The Voyage of the Beagle* owed something – quite a lot, in fact – to Darwin's process of putting evolution to bed while he tackled barnacles.

It is easy to identify the new material in the 1845 edition

of *The Voyage of the Beagle*, once it has been pointed out. By comparing the first and second editions, you can pull out all the new material and put it together, so that, in Gruber's words, 'taken out of their hiding places and strung together' these paragraphs form 'an essay which gives almost the whole of [Darwin's] thought' on evolution by natural selection, without actually spelling out the law of natural selection itself. One of the key passages, for example, is the addition of 'a very clear statement of the Malthusian principle of the relation between food supply and population growth', used to account for the 'increasing rarity and eventual extinction of some species'. Another refers to the variety of finches found in the Galapagos Islands. At least one person, John Lindley, the editor of *The Gardeners' Chronicle*, noticed the changes and drew attention to some of them in the *Chronicle*; in response, Darwin wrote to Lyell, 'I was much pleased by Lindley picking out my extinction paragraphs and giving them uncurtailed.' The only explanation for all this is that Darwin was concerned about posterity, and about his priority. If anyone else came up with the idea, he would be able to point to this 'ghost' essay and reveal that he had thought of it first.

As we have seen, Darwin picked up the threads of his evolutionary thinking and started to write his 'Species Sketch' in May 1856, fired on by Lyell's concerns that he might be pre-empted. This was a major undertaking. In November 1856 he wrote to Lyell:

I am working very steadily at my big book; I have found it quite impossible to publish any preliminary essay or sketch; but am doing my work as completely as present materials allow without waiting to perfect them. And so much acceleration I owe to you.

By June 1858 he had produced ten chapters (not quite one every two months) and was about two-thirds of the way through the book, which was intended as a scientific tome for the cognoscenti, not in any sense a popularisation of his ideas aimed at the same audience as the *Vestiges*. Everything must have seemed set fair for his theory to be unleashed on the scientific community in another year or two. Then Wallace's bombshell arrived. Darwin's first reaction was to tell Lyell what had happened. He wrote, enclosing Wallace's essay:

> *Your words have come true with a vengeance – that I should be forestalled. You said this, when I explained to you here very briefly my views of 'Natural Selection' depending on the struggle for existence. – I never saw a more striking coincidence; if Wallace had my M.S. sketch written out in 1842, he could not have made a better short abstract! Even his terms now stand as Heads of my Chapters.*
>
> *Please return me the M.S. which he does not say he wishes me to publish; but I shall of course at once write & offer to send to any Journal. So all my originality, whatever it may amount to, will be smashed. Though my Book, if it will ever have any value, will not be deteriorated; as all the labour consists in the application of the theory. I hope you will approve of Wallace's sketch, that I may tell him what you say.*

Lyell, however, was not convinced that Darwin should give up his priority so easily. Wallace's essay was also shown to Darwin's friend Joseph Hooker (1817 to 1911),* and Lyell and Hooker discussed what to do next; Darwin was largely a bystander to

---

* Hooker, eight years younger than Darwin, was a leading naturalist and in 1865 succeeded his father as Director of Kew Gardens. He also features in *Flower Hunters*.

these discussions, preoccupied with the illness of his son Charles Waring Darwin, who died from scarlet fever on 28 June, at the age of eighteen months. He also agonised about whether he had any right to claim priority, writing to Lyell:

> *Wallace says nothing about publication. But as I had not intended to publish any sketch, can I do so honourably because Wallace has sent me an outline of his doctrine? I would far rather burn my whole book, than that he or any other man should think that I had behaved in a paltry spirit.*

It seems to have been Hooker who came up with the solution to the problem, taking advantage of a meeting of the Linnean Society which should have taken place on 17 June but had been adjourned to 1 July as a mark of respect to a former president of the society, who had just died. Making free use of the material supplied by Darwin and without him being aware of any of the details, Lyell and Hooker arranged to present to that meeting, in what they regarded as the appropriate chronological order of writing, an abstract from Darwin's 1844 sketch, part of a letter that Darwin had written to Asa Gray, in Boston, in 1857, and Wallace's essay. Darwin's contribution amounted to about 2,800 words; Wallace's to some 4,200 words. In the Proceedings of the Linnean Society this appeared as a joint paper with the title and authorship 'On the tendency of species to form varieties; and on the perpetuation of varieties and species by means of natural selection' by Charles Darwin Esq., FRS, FLS, & FGS and Alfred Wallace Esq., communicated by Sir Charles Lyell, FRS, FLS, and J. D. Hooker Esq., MD, VPRS, FLS, &c.'*

---

* *Journal of the proceedings of the Linnean Society*, Zoology II, 1858, page 45. Unfortunately, after the *Journal* went to press the printer, as was the usual practice, threw away the manuscripts of its articles, including Wallace's Ternate essay. Wallace had not made a copy.

The 'joint paper' did not cause a stir either at the meeting or when it was published. In his Presidential Report to the Linnean in May 1859, looking back at 1858, Thomas Bell said, 'The year which has passed has not, indeed, been marked by any of those striking discoveries which at once revolutionise, so to speak, the department of science on which they bear.' And in his autobiography Darwin recalled 'our joint productions excited very little attention, and the only published notice of them which I can remember was by Professor Haughton of Dublin, whose verdict was that all that was new in them was false, and what was true was old.' The publication that did make an impact, and which put natural selection in the public eye, was not the joint paper, but Darwin's book. This is relevant to the questions sometimes raised about whether Wallace, far away on the other side of the world and without a voice in the debate, was treated fairly in these machinations.

The conspiracy theorists make much of the fact that in their introduction to the joint paper Lyell and Hooker state 'both authors having now unreservedly placed their papers in our hands.' How could Wallace have given permission for anything in the fortnight between his essay reaching Down House and the meeting of the Linnean? But this is to misunderstand both the tradition of the day and the literal meaning of 'unreservedly'. It was the usual practice, as we have seen both with Darwin's letters from the *Beagle* voyage and Wallace's letters from South America, for items of scientific interest to be taken and published as soon as possible. If a communication was not intended for possible publication or wider dissemination the writer would mark it 'private', or might pick out a particular passage with the instruction that it should not be shown to anyone else. Darwin's 1857 letter to Asa Gray is a good example; on that occasion he specifically asked Gray not to divulge the details of his theory.

Wallace included no such reservations when he wrote to Darwin with his essay; he sent it 'unreservedly', in the full knowledge that it might (indeed, he hoped it would) be shown to others. Having it published was more than he expected, but something he appreciated. When he heard the news, he wrote to his mother expressing his pleasure:

> *I have received letters from Mr. Darwin and Dr. Hooker, two of the most eminent naturalists in England, which have highly gratified me. I sent Mr. Darwin an essay on a subject upon which he is now writing a great work. He showed it to Dr. Hooker and Sir Charles Lyell, who thought so highly of it that they had it read before the Linnean Society. This insures me the acquaintance of these eminent men on my return home.*

Wallace was acutely aware that having his name linked with those of Lyell, Darwin and Hooker enhanced his prestige, and drew attention to his work. He also wrote to Hooker from Ternate on 6 October 1858 (the same day that he wrote to his mother):

> *Allow me in the first place sincerely to thank yourself and Sir Charles Lyell for your kind offices on this occasion, and to assure you of the gratification afforded me both by the course you have pursued, and the favourable opinions of my essay which you so kindly expressed. I cannot but consider myself a favoured party in this matter, because it has hitherto been too much the practice in cases of this sort to impute all merit to the first discoverer of a new fact or a new theory, and little or none to any other party who may, quite independently, have arrived at the same result a few years or a few hours later.*

Throughout the rest of his life he never missed an opportunity to express his gratitude; this comment, from 1903, is typical:

> *My connection with Darwin and his great work has helped to secure for my own writings on the same questions full recognition by the press and the public; while my share in the origination and establishment of the theory of Natural Selection has usually been exaggerated.* *

But if Wallace had achieved nothing more than forcing Darwin to write the *Origin*, it would have been a major contribution to science. Wallace commented on this in *My Life*:

> *Darwin [later] wrote that he owed much to me and his two friends, adding: 'I almost think that Lyell would have proved right, and that I should never have completed my larger work.' I think, therefore, that I may have the satisfaction of knowing that by writing my article and sending it to Darwin, I was the unconscious means of leading him to concentrate on the task of drawing up what he termed an 'abstract' of the great work he had in preparation, but which was really a large and carefully written volume.*

Judging from the (lack of) response to the joint paper, even with the weight of Lyell and Hooker behind it, and the earlier lack of response to Wells and Matthew, the theory of natural selection would indeed have continued to languish without Darwin's book. But even Darwin, usually so slow and careful

---

* 'My relations with Darwin in reference to the theory of natural selection', *Black and White*, 17 January 1903.

about his writing, now realised that it was time for speed and, by his standards, brevity.

After the funeral of Charles Waring, the Darwins left Down House to get away from all the trauma. They arrived at Sandown, on the Isle of Wight, on 17 July. At that time, Darwin's intention was to write a summary of his theory for publication in the *Journal* of the Linnean, but as he wrote to Hooker on 13 July, 'How on earth I shall make anything of an abstract in 30 pages of Journal I know not'. He quickly realised that this would be impossible, and instead set to work converting the material he already had in hand for his intended big book into something shorter and more accessible. In correspondence, Darwin referred to this as a 'small volume' and he still thought of it as an 'abstract' of his full theory; but it became *On the Origin of Species*. The 'big book' was never published in the originally anticipated form, but a lot of the material left out of the *Origin* appeared in other places, notably in a two-volume work on *Variation* published in 1868. The writing of the *Origin* itself was completed on 19 March 1859, shortly after Darwin's fiftieth birthday. It ran to 155,000 words – about twice the length of the present book. On Lyell's advice, he sent the manuscript to the publisher John Murray, and it was in the bookshops on 24 November, with the impressive title *On the Origin of Species by Means of Natural Selection, or the Preservation of Favoured Races in the Struggle for Life*. Most accounts mention that the first print run of 1,250 copies was sold out on the day of publication, but this is only true in the sense that all the copies had been bought up by the bookshops, ready to sell on to their customers. Nevertheless, the book was an immediate success, and its publication did indeed mark the moment when the idea of evolution by natural selection became part of mainstream science and public debate. The reaction of many of Darwin's contemporaries is probably best understood by the famous remark of Thomas Henry Huxley, who, looking back on the events nearly thirty years later, wrote:

> *I imagine that most of those of my contemporaries who thought seriously about the matter, were very much in my own state of mind – inclined to say to both Mosaists and Evolutionists, 'a plague on both your houses!' and disposed to turn aside from an interminable and apparently fruitless discussion, to labour in the fertile fields of ascertainable fact. And I may, therefore, further suppose that the publication of the Darwin and Wallace papers in 1858, and still more that of the 'Origin' in 1859, had the effect upon them of the flash of light, which to a man who has lost himself in a dark night, suddenly reveals a road which, whether it takes him straight home or not, certainly goes his way . . . The 'Origin' provided us with the working hypothesis we sought. Moreover, it did the immense service of freeing us for ever from the dilemma – refuse to accept the creation hypothesis, and what have you to propose that can be accepted by any cautious reasoner? In 1857 I had no answer ready, and I do not think that anyone else had. A year later we reproached ourselves with dullness for being perplexed by such an inquiry. My reflection, when I first made myself master of the central idea of the 'Origin' was, 'How extremely stupid not to have thought of that!' . . . Darwin and Wallace dispelled the darkness, and the beacon-fire of the 'Origin' guided the benighted.*[30]

Huxley, who had been born in 1825 and died in 1895, was a towering figure in the biological sciences in the second half of the nineteenth century, and became a leading figure in the debate that followed the publication of the *Origin*, speaking out in support of the theory when Darwin was too ill or too reticent to join in public discussion, and becoming known as 'Darwin's bulldog'. But as well as filling in for Darwin, Huxley was in effect also filling in for Wallace, who could only follow events back home with a delay of several weeks, when the mail steamers arrived.

The letters from Darwin and Hooker that prompted Wallace's letters to his mother and Hooker on 6 October 1858 are lost, but Wallace copied out an intriguing extract from Darwin's letter in his notebook. This is a list of the subjects to be covered in the fourteen chapters of Darwin's planned big book, with the comment that he had completed everything up to Chapter Ten, when the Ternate essay arrived. This is the only surviving record of Darwin's intentions regarding this publication. Darwin also sent Wallace one of the first copies of the *Origin* – he may even have sent him a set of proofs. In the light of all this, Wallace quietly dropped his plans to write a book of his own on natural selection, and concentrated on collecting for the remainder of his time in the East.

Meanwhile, back home the debate about evolution was stirred up by Darwin's book, even though the idea of natural selection proved hard to sell. The problem was heredity – nobody knew how characteristics could be passed on from one generation to the next, nor how subtle changes could be introduced during this process. So people still sought other mechanisms, and even Darwin, responding to criticism, included a role for a modified form of Lamarckism in later editions of the *Origin*, so that the first edition actually provides the clearest explanation of his ideas. The most fervent public supporter of natural selection over the next couple of decades would be Wallace, not Darwin. And, intriguingly, some of the best evidence in support of natural selection came from Wallace's former travelling companion, Henry Bates.

While studying and classifying the butterflies he had collected in the Amazon basin, Bates realised that in some cases butterflies that carried the distinctive markings of a poisonous species were not themselves poisonous, but were members of a different species. Just as predators had somehow learned to avoid eating the poisonous butterflies, so the mimics had somehow developed the same markings, which also protected them from predators. The 'somehow', of course, we now understand to be evolution by natural selection. Predators with a liking for the poisonous butterflies die, and do not pass on their liking to later generations; predators that avoid those butterflies survive, and the propensity to avoid them is passed on. So any butterfly that resembles the poisonous ones is more likely to survive, and over many generations this results in mimics more and more like the species they resemble, in a way reminiscent of the way evolution of the tiger beetles that fascinated Wallace had produced species that matched the background they lived on. The process has become known as Batesian mimicry. Not long after he returned to England from South America, Bates described his findings at a meeting of the Linnean on 21 November 1861; the paper was published the following year in their *Transactions*, and he elaborated on the theme in his book *The Naturalist on the River Amazons*, published by Murray in 1863. But even this evidence did not tilt the balance of opinion, even amongst those who now accepted the reality of evolution, in favour of natural selection.

Darwin had actually put natural selection at the front of the *Origin*, because he knew that what the idea of evolution had lacked was a mechanism. It was only later in the book that he marshalled the evidence for evolution, elaborating on the geographical distribution of living species, the fossil record, variation under domestication and evidence from comparative anatomy. One of his most powerful images was the 'branching tree' analogy that had also occurred to Wallace:

> *The green and budding twigs may represent existing species; and those produced during each former year may represent the long succession of extinct species. At each period of growth all the growing twigs tried to branch out on all sides, and so overtop and kill the surrounding twigs and branches, in the same manner as species and groups of species have tried to overmaster other species in the great battle for life.*

He also explained that 'primitive' species could survive unchanged for a very long time when they were well suited to an unchanging environment, while even more 'advanced' organisms would go extinct when the environment changed. And he spelled out what he meant by the 'struggle for existence':

> *I use [that term] in a large and metaphorical sense, including dependence of one being on another, and including (which is more important) not only the life of the individual, but success in leaving progeny.*

What would become an uncomfortable topic of debate in the decades ahead, as we mentioned in Chapter Three, was also highlighted with his comment 'the lapse of time has been so great as to be utterly inappreciable by the human intellect.'

The most important message that he got across in the book, however, was the idea that modern species share a common descent from some single individual: 'probably all the organic beings which have ever lived on this earth have descended from some one primordial ancestor'. Offering as a sop to any religiously minded readers only the word 'breathed', he summed up:

Thus, from the war of nature, from famine and death, the most exalted object which we are capable of conceiving, namely the production of the higher animals, directly follows. There is a grandeur in this view of life, with its several powers, having been originally breathed into a few forms or into one; and that, whilst this planet has gone cycling on according to the fixed law of gravity, from so simple a beginning endless forms most beautiful and most wonderful have been, and are being, evolved.

After the publication of the *Origin*, there were still intense debates among scientists about the mechanism of evolution. But the fact of evolution became accepted – not overnight, but over a decade or so. The fiercest arguments now raged around humankind's place in evolution, a subject addressed by both Darwin and Wallace.

Wallace left Singapore on 8 February 1862, just after his thirty-ninth birthday, and was back in England on 31 March, having persuaded the Zoological Society to pay for his First Class passage as part of a deal for him to bring back two live Birds of Paradise. The profits from his expedition had been invested wisely by his agent, and the investments were bringing in £300 a year, enough for a single man without obligations to live on comfortably. Unfortunately (from a purely financial perspective), Wallace did not remain single and had other obligations. More on that shortly.

Even before he got back to England, on 19 March Wallace was elected as a Fellow of the Zoological Society, and his return was eagerly awaited by his fellow naturalists. He was able at last to meet Charles Lyell; but he had to postpone until later in the year an immediate invitation to visit Darwin, because he was laid up with a variety of relatively minor ailments resulting from the hard life he had been living. At least this gave him plenty of time to catch up with his reading and get up to date with the evolution debate. He recuperated at the house of his sister Fanny and her husband, who were now living in Paddington, where Thomas Sims was running a not very successful photographic business with his brother Edward. Wallace had a huge amount of material to sort out from his travels, including selling his latest collections, which arrived after him having followed the cheaper route round the Cape. He was also immediately involved in supporting his family. He sent money to his mother, paid the Sims's rent, and over several years provided £700 to keep the photographic business going. In spite of the success of his expedition, sooner or later he would, after all, still need to find some source of income.

On a happier note, he renewed acquaintance with George Silk, now living in Kensington, and Bates, returned from the Amazon and highly regarded for his paper on mimicry. While he was living in London, Wallace also had opportunities to meet up with Darwin when Charles came to stay with his brother Erasmus. 'On these occasions,' Wallace writes in *My Life*, 'I usually lunched with him and his brother, and sometimes one other visitor, and had a little talk on some of the matters specially interesting him.'

All of this delayed the writing of Wallace's book on *The Malay Archipelago*, which eventually appeared in 1869. But starting in 1864 he produced a stream of scientific papers and articles based on his work in the region. He was also busy, in a quiet way, with a more personal project. He was eager to settle into the kind of comfortable domestic life that he saw people like Darwin enjoying,* and he became attracted

---

* In his autobiography he wrote, 'if the entire proceeds of my Malayan collections had been well invested, and I had obtained a secure income of £400 or £500 a year, I think it probable that I should not have written another book, but should have gone to live further in the country, enjoyed my garden and greenhouse . . .'

to Marion Leslie, the daughter of a friend, Lewis Leslie; she was then in her late twenties. After some reluctance on her part, they became engaged; but in 1864 she had a change of heart and the engagement was broken off. By this time Richard Spruce was back in England and dividing his time between London and the village of Hurstpierpoint, in Sussex, where Wallace visited him in the autumn in the aftermath of the broken engagement. There, he met Spruce's friend William Mitten, a pharmacist and amateur naturalist, his wife and their four daughters. A year and a half later, in April 1866, Wallace married the eldest daughter, Annie, then twenty years old. She soon became pregnant, increasing the need for Wallace to find a regular source of income.

Partly through financial necessity, and largely through the encouragement of his wife, Wallace settled down and concentrated on writing his book. Unfortunately, in an effort to increase his income he also began speculating on the stock market, which had the opposite effect from what he had hoped. On 22 June 1867, his son Herbert Spencer Wallace was born, and the family moved to Hurstpierpoint to be close to Annie's family while he worked on his writing. Spruce was also there, and this seems to have been one of the happiest periods of Wallace's life. In November 1868 (the same month that his mother died) he was awarded the Royal Medal of the Royal Society, on 27 January 1869 his daughter Violet was born, and on 9 March *The Malay Archipelago* was published, with an advance of £100 from Macmillan publishers and the promise of a royalty on every copy sold after the first thousand. Now, it was time to find a paying position.*

Meanwhile, Darwin had also been busy writing. Although suffering recurring bouts of severe illness, he produced *On*

---

* From now until the end of his life Wallace moved house several times, but there is no need to go into all the details.

the Various Contrivances by which British and Foreign Orchids are Fertilised by Insects, published in 1862, his big book on *The Variation of Animals and Plants under Domestication* (1868), incorporating much of the material left out of the *Origin*, and in 1871 the book most relevant to our story, *The Descent of Man*. This was far more of an immediate best-seller than the *Origin*, with 4,500 copies in print within two months of publication. The significance of the book can be summed up from the opening words of an essay that Darwin wrote in 1839, twenty years before the publication of the *Origin*, but never published:

> *Looking at Man, as a Naturalist would at any other Mammiferous mammal . . .*[31]

That was the heart of the *Descent*, in which Darwin was indeed 'Looking at Man, as a Naturalist would at any other Mammiferous mammal . . .' and argued the case that we have formed by the same process of evolution by natural selection as all other species. This provoked another public debate, with the main proponent of the idea that humankind has a special place in creation now being St George Jackson Mivart (1827 to 1900), a devout Roman Catholic who was also a biologist and a Fellow of both the Linnean and Royal Societies. His book *On the Genesis of Species* was also published in 1871. His key objection to the Darwin-Wallace theory was that evolution could not have proceeded in tiny steps because, for example, a creature with a neck longer than that of a deer but shorter than that of a giraffe would have no evolutionary advantage because it could not browse on tree tops. So, he argued, evolution must proceed in jumps – saltations – with, in effect, a deer giving birth to a giraffe, specially created to fit a new ecological niche. The same argument is still trotted out today, usually by asking the question 'what use is half an eye?'; for the answer, we recommend Richard Dawkins's book *The Blind Watchmaker*.[32] Mivart also required supernatural forces to be at work to produce the human 'soul'; but leaving aside his religious objections, the relevant point is one of timescale. Any species can be turned into any other species by a number of tiny steps if there is sufficient time; but the availability of a sufficiency of time did not become clear until the revolution in physics in the twentieth century.

In 1871 Darwin passed his sixty-second birthday. He spent the last decade of his life working on many projects, including revising both the *Origin* and the *Descent*, a book on *The Expression of the Emotions in Man and Animals* (published in 1872, it was one of the first books to contain photographs), *Insectiverous Plants* (1875), *The Effects of Cross and Self Fertilisation in the Vegetable Kingdom* (1876), a substantially enlarged version of *Fertilisation of Orchids* (1877), *The Different Forms of Flowers on Plants of the Same Species* (1877), and last, but far from least, one of his most enjoyable books, *The Formation of Vegetable Mould, through the Action of Worms, with Observations on their Habits* (1881). He was able to achieve all this because he never had to worry about money, and never got involved in the day-to-day running of things like the Linnean and Royal Societies; he was not a committee man. When he died in 1882, he was carried to his grave by one of the most illustrious groups of pall-bearers ever assembled – they included two knights, Sir Joseph Hooker and Sir John Lubbock, two Dukes, Argyll and Devonshire, the Earl of Derby, the President of the Royal Society, Thomas Henry Huxley . . . and Alfred Russel Wallace.

The contrast between Darwin's final years and Wallace's contemporary life could hardly have been starker. After the publication of *The Malay Archipelago*, which he dedicated to Charles Darwin, Wallace's scientific reputation should have been assured. But in spite of applying for many posts, including Assistant Secretary of the Royal Geographical Society, he was never successful and had to depend on the erratic income from his writing and chores such as marking exam papers. The rejections were in part influenced by the fact that by the end of the 1860s Wallace had become an ardent and outspoken spiritualist. Even though this was something of a fad at the time, it was a distinct aberration for a serious scientist. Ironically, though, as we shall see, it was a connection developed through his interest in spiritualism that eventually ensured that Wallace lived in relative comfort in his old age. These beliefs encouraged Wallace to think that humankind had not evolved entirely through the same process that had guided the evolution of other species. At the end of *The Malay Archipelago* he wrote:

> We most of us believe that we, the higher races, have progressed and are progressing. If so, there must be some state of perfection, some ultimate goal, which we may never reach, but to which all true progress must bring us nearer.

This was just a taste of things to come. In the April 1869 edition of the *Quarterly Review* he wrote:

> The moral and higher intellectual nature of man is as unique a phenomenon as was conscious life on its first appearance in the world, and the one is almost as difficult to conceive as originating by any law of evolution as the other . . . an Overriding Intelligence has watched over the action of those laws, so directing their variations and so determining their accumulation as finally to produce . . . the indefinite advancement of our mental and mortal nature.

On his copy of the *Quarterly Review*, Darwin wrote 'No', underlined three times, alongside these words; and in a letter to Wallace, he wrote, 'I differ grievously from you'.

Early in 1870, Wallace published, to some acclaim from his peers, a collection of his articles under the title *Contributions to the Theory of Natural Selection*. But around the same time he also got embroiled in an argument which would, really through no fault of his own, adversely affect his reputation. A 'flat Earther' called John Hampden issued a challenge to the scientific community 'to exhibit, to the satisfaction of any intelligent referee, a convex railway, river, canal or lake' and offered a bet of £500 on the result. Either because of the financial lure, or in an effort to defend science (or both), Wallace took up the bet, although he first took the precaution of asking Lyell for his advice on whether to do so. Lyell's reply, according to Wallace, was to go ahead because 'it may stop these foolish people to have it plainly shown them.'[33] Wallace devised a very simple experiment which took place along a six-mile stretch of the Bedford Canal. It's worth going into a few details, since there are still foolish people around who claim not to believe that the Earth is round. At each end of the stretch of water, Wallace erected markers the same height above the level surface of the water. In the middle, there was another marker, also at the same height above water. Using his surveying skills, Wallace could sight along the line of the markers from one end to the other. If the Earth were flat, the marker in the middle would be exactly along the line of sight. But because of the curvature of the Earth it was actually lifted up above the line of sight. The evidence was accepted by 'an intelligent referee' approved by both parties – the editor of *The Field* – and the results published in his journal. But when Wallace claimed his reward, Hampden refused to pay up. It might have been wiser to leave it there, but Wallace tried to make

Hampden live up to his promise, and got involved in legal wrangling which went on for about two decades and cost him money. Hampden, clearly unhinged, took to writing derogatory letters about Wallace to all the learned societies, and even to Mrs Wallace. This may also have affected Wallace's prospects of employment – even though he was in the right, he was perceived as behaving in an unseemly manner.

More happily, Wallace's big scientific project in the early 1870s was a well-received two-volume book on *The Geographical Distribution of Animals*, which appeared in 1876. But he continued to take issue with Darwin concerning humankind's place in nature, and when he reviewed the *Descent* he commented that the:

> . . . *absolute erectness of [Man's] posture, the completeness of his nudity, the harmonious perfection of his hands, the almost infinite capacities of his brain, constitute a series of correlated advances too great to be accounted for by the struggle for existence of an isolated group of apes in a limited area.*

Wallace's last child, William, was born on 30 December 1871 (his first, Bertie, would die in 1874, aged six), and in March 1872 he was elected (somewhat belatedly, you might think) as a Fellow of the Linnean Society. His next project was another large book, *Island Life*, published in 1880. By the time it appeared in print, Wallace's financial situation was worse than ever; he was saved by a spiritualist connection, in spite of reservations related to some of his activities.

Wallace had become a close friend of Arabella Buckley, a fellow spiritualist who had been the secretary of Charles Lyell. Although Lyell had died in 1875, Buckley knew all the great men of science of the day, and she was also well aware of Wallace's financial struggles. At the end of 1879 she wrote to Darwin to ask if he could use his influence to get Wallace a post, no matter how modest. Darwin was all in favour and wrote to Hooker to enlist his support, perhaps in obtaining for Wallace a government pension. Hooker's reply was caustic. Wallace, he said, had 'lost caste terribly' because of his 'outspoken support of spiritualism' and by 'taking up the Lunatic bet about the sphericity of the earth'. Besides, Wallace was not 'in absolute poverty' and therefore not worthy of a pension. Taken aback, Darwin told Buckley that the situation was hopeless. In blissful ignorance of all this, Wallace dedicated *Island Life* to Hooker, 'who, more than any other writer, has advanced our knowledge of the geographical distribution of plants, and especially of insular flora', and sent him a copy on publication in November 1880.

By then, Darwin was having second thoughts. After all, Hooker wasn't the only influential scientist around. He sounded out his neighbour, the anthropologist John Lubbock, and Huxley, who offered to try to get Hooker on board. Darwin then asked Buckley for some background material on Wallace to use in drawing up a petition to the government. The timing was perfect. Hooker was impressed by *Island Life* (not just the dedication) and changed his tune. With Hooker on board, they soon gathered an impressive list of supporters. A petition – or 'memorial', in the language of the day – was signed by dignitaries including the President of the Royal Society, the President of the Linnean Society, the Director of the Geological Survey, Lubbock, Bates, Hooker, Huxley and Darwin, and presented to the Prime Minister, William Gladstone. The result was that Wallace was awarded a pension of £200 per year, backdated to July 1880. The news reached him on his fifty-eighth birthday. Although not enough for a life of luxury, the income ensured that he was never in real hardship.

Following Darwin's death – if not before – Wallace was the leading proponent of and spokesman for the theory of evolution by natural selection, which he always referred to as 'Darwinism'. Even Darwin had backed away from his original position, concerned by the twin objections of the timescale required for evolution and the lack of a satisfactory mechanism to explain heredity, and willing even to embrace a modified form of Lamarckism. But Wallace, ironically in view of his disagreement with Darwin about humankind's place in the evolutionary scheme, remained a pure natural selectionist, more Darwinian than Darwin himself. In this capacity he made a successful ten-month lecture tour of the United States and Canada in 1886 and 1887, and used the material from his lectures as the basis for his book *Darwinism*, published in 1889. This was an important and timely overview of the theory, in an era when it was under fire for the reasons we have already mentioned, and is still well worth reading.

By then in his late sixties, Wallace had become one of the Grand Old Men of Victorian science, and he lived until 7 November 1913, when he was nearly ninety-one years old. He continued to write, and received many honours, culminating in the Order of Merit, Britain's highest civilian honour, in 1908. But more significantly, from our point of view, he lived long enough to see at least the beginnings of the answers to the two questions that had bothered Darwin so much, the timescale problem and the heredity problem. The resolution of the timescale problem was discussed in Chapter Three; unknown to either Darwin or Wallace, however, the first clues to the resolution of the heredity problem had already been discovered in the 1860s, when Darwin was working on the *Variation*, and Wallace was writing up his material from the

East, culminating in *The Malay Archipelago*. But telling the rest of the story of evolution requires a change both of pace and focus from the relatively leisurely and broad-brush approach of the Victorian era.

PART THREE

MODERN TIMES

CHAPTER SEVEN

FROM WRINKLY PEAS TO CHROMOSOMES

In the twentieth century, scientific progress proceeded more rapidly than ever before, and the understanding of evolution at work began to focus less on whole animals and plants and more on what goes on inside the cells of animals and plants. That is where the key to an understanding of the mechanism of heredity lay. Over the same interval of time, the study of evolution changed from being primarily concerned with observations of the behaviour of the living world to being primarily about experiments. Sometimes, though, the significance of experiments is not widely appreciated at first, either because they receive little publicity, or because they do not fit into the framework of current thinking – or both, as in the case of Gregor Mendel's investigation of inheritance in peas.

The key to understanding evolution, a realisation made by

both Darwin and Wallace, is that like begets like, but imperfectly. The offspring of a male and female cat will always be cats, not canaries or codfish, or willow trees. There are no 'hopeful monsters'. But none of their offspring will be an exact copy of either parent. The hereditary mechanism for this imperfect copying baffled Darwin, although he made repeated attempts to tackle the puzzle in the 1860s and 1870s.

Darwin's (incorrect) model of heredity was initially put forward in a self-contained chapter at the end of his book on *Variation*, in 1868, and elaborated at various times, including in later editions of the *Origin* (one reason why the first edition is better than its successors). He gave this idea the name 'pangenesis'; 'pan' from the Greek meaning 'all', because Darwin thought that all the cells in a body are involved, and 'genesis', of course, for reproduction. The essence of his idea, which he described in *Variation* as 'a provisional hypothesis or speculation', is that every cell in the body produces tiny particles called 'gemmules' which travel to the reproductive cells (egg or sperm) and are passed on to the next generation. This contained an element of Lamarckism, because the production of gemmules could be affected by the environment – we might imagine that if the climate got colder the gemmules might be affected to encourage the growth of fur in later generations. But, like many of his contemporaries, Darwin also thought of inheritance as somehow producing a blending of the characteristics of each parent. In a simple example, blending would imply that the offspring of a blond man and a dark-haired woman would all have brown hair. Such a situation would be very bad for evolution, because it would iron out the differences among individuals on which natural selection operates – Wallace's tiger beetles, for example, could never achieve a perfect match with their background. In the real world, the offspring of one fair parent and one dark parent may themselves be either dark or fair. Or they may turn out to have ginger hair, unlike either parent. It was this aspect of inheritance that Mendel's experiments, carried out and even published while Darwin was still alive, explained. But his discovery remained largely unknown until the beginning of the twentieth century.

Mendel was born in 1822, six months before Wallace, and lived until 1884. He came from a poor family in a small hamlet in what was then Moravia, a region covering the borders of modern Poland, Germany and the Czech Republic. He was christened Johann, and thrived at school, but the only congenial occupation open to a clever young man with his background was the priesthood. In 1843 he was admitted as a novice of the Augustinian order based at Brünn (now Brno) and took the name Gregor. Rising through the ranks of the priesthood, Mendel became a schoolteacher and was sent by his Prelate to study at the University of Vienna from 1851 to 1853. He was not 'just' a priest, but also a trained scientist. This was not unusual; the monastery at Brünn was a kind of mini-university, not just a religious centre, and included a botanist and an astronomer among its members. Although Mendel's main role in the community was as a teacher at the local school, and he had his religious duties to carry out, he was also allowed time to carry out his own experiments on the way heredity works. He had become fascinated by the way characteristics are passed on from one generation to the next, and started out by breeding mice, but in 1856 he turned to botany for what became his landmark work, involving pea plants.

Mendel chose peas for good reasons, after investigating several other plants. He knew that they had distinctive characteristics that bred true and which could be analysed statistically. The statistical analysis was the key to his work, and far ahead of his time. He picked out several characteristics to study, such as whether the seeds were wrinkly or smooth, whether they were yellow or green. His uniquely original contribution was that he approached the study of biology like a physicist. He carried out repeatable experiments, kept detailed records, and used statistical tests to analyse the data. From an initial 28,000 plants, he selected 12,835 for detailed study. For each plant Mendel kept a record of its descendants, like a family tree. He knew the parents, grandparents, and even earlier ancestors of each plant in succeeding generations. This was only possible because he fertilised every flower of each of the thousands of plants by hand, dusting the pollen from a specific single plant onto the flowers of another specific single plant. As the plants developed, he had to note the relevant characteristics of each individual while tending the crop, then repeat the whole process in subsequent generations. It took seven years to build up a database that allowed him to work out how the characteristics he was studying were passed on from one generation to another.

Just one example, the inheritance of wrinkly (or rough) or smooth seeds, highlights what he found. Mendel found that something in a plant is passed from one generation to the next and determines the nature of the offspring. We now call that something a gene, or a package of genes; Mendel did not use that term, instead referring to 'hereditary elements', and they are also known as 'factors'; but we shall stick with the modern terminology. His statistical analysis showed that the properties he studied related to pairs of genes. In our example, one gene is associated with roughness (R), and one gene with smoothness (S). Each individual plant inherits one possibility from each parent. As a result, the offspring may possess any one (but only one) of the combinations RR, RS or SS. It passes one of these possibilities on to the next generation. An RR or SS plant must pass on R or S, respectively, but an RS plant will pass on R to half its offspring and S to the other half. RR plants, Mendel found, always have rough seeds. SS plants always have smooth seeds. But in RS plants, according to Mendel's meticulous statistical analysis, the R is ignored and the peas are all smooth.

He found this by crossing plants that always produce rough seeds (RR) with plants that always produce smooth seeds (SS). Just 25 per cent of the offspring had rough seeds, and 75 per cent had smooth seeds. Mendel explained that this can only be because although 25 per cent of the offspring are RR, and 25 per cent are SS, producing rough or smooth seeds respectively, the rest are 25 per cent RS and 25 per cent SR, adding up to 50 per cent, and both producing smooth seeds. Crucially, the RS and SR plants do *not* produce 50 per cent R seeds and 50 per cent S seeds, nor seeds that are just a little bit rough. We now say that the S factor is dominant, and the R factor is recessive.

Mendel's results were presented to the Brünn Society for the Study of Natural Science in February 1865 and published in their proceedings in 1866, but this was even then an obscure journal and their significance was not appreciated. The combination of botany and mathematics seems to have baffled the few people who read the paper, although it seems natural today. Mendel became Abbot of his community in 1868, and had no time for further scientific research. It was only at the end of the nineteenth century, when other researchers independently discovered the same laws of inheritance, that his papers were rediscovered and he was given the credit he deserves. The five key points that he highlighted are:

- Each physical character of an organism corresponds to one hereditary factor.
- Factors come in pairs.
- One, but only one, factor from each pair is passed on by each parent to its offspring.

- There is an equal probability, in a strict statistical sense, of either factor of a pair being passed on in this way to any individual offspring.
- Some factors are dominant and others are recessive.

The laws of inheritance that Mendel discovered are of key importance in understanding the theory of evolution by natural selection. First, they explain why offspring do not have properties that are a blend of the characteristics of their parents. Second, Mendel showed that each characteristic is inherited independently. Whether or not the pea is green or yellow, for example, does not affect whether it is rough or smooth. The next step towards an understanding of the mechanism of evolution would be taken early in the twentieth century, by Thomas Hunt Morgan (1866 to 1945). But to put that in context, we need to backtrack a little to the identification of cells as the basic units of life.

The name 'cell' was first used in a biological context by Robert Hooke, to describe structures he saw when he studied slices of cork under the microscope. These reminded him of the tiny rooms, or cellula, occupied by monks. The structures we now call cells are even smaller than the ones Hooke studied, but when nineteenth-century biologists probed the structure of living matter with improved microscopes, they took over the name. It was only in 1838 that Matthias Schleiden, a German botanist who lived from 1804 to 1881, suggested that all plant tissues are made of cells, and a year later his compatriot Theodor Schwann (1810 to 1882) proposed that all forms of life – animal as well as plant – are based upon cells. In the 1840s they developed the idea that cells are the basic units of life, and pointed out that not only do individual cells possess all the attributes of life, but all of the complexity of larger organisms is built from an underlying structure of cells. For the first time, it became appreciated that an egg or a seed are individual cells that are capable of reproducing, dividing to produce more cells which become organised into the mature form of an organism. As Schleiden put it, the organism is a 'cellular state', in which 'each cell is a citizen'.[34] Previously, life had been regarded as some mysterious property of a whole organism; now, it was seen as a property shared by even the humblest cells.

This led to another profound realisation. At the end of the 1850s, studies by another German, Rudolf Virchow (1821 to 1902), building on the work of Robert Remak (1815 to 1865), showed that no cell ever came into existence spontaneously.* Wherever there is a cell, he pointed out in 1858 (the year of the Darwin–Wallace joint paper), there must have been a previous cell. Just as animals always have parents and plants are only produced from the seeds of other plants, cells are produced only by the division of other cells. Life never appears spontaneously on Earth today. All living cells are descended, in an unbroken line, from some remote ancestor (or ancestors) in the distant geological past. Although Virchow did not quite go so far as to suggest that there was literally one single cell that is the ancestor of all life on Earth today, this is now widely accepted as the most likely explanation of the similarity of all life on Earth at a molecular level. The origin of the first cell remained a mystery; but following Virchow's work there was no mystery about the origin of the life in animals and plants today.

Once all this was fully appreciated, the study of life became the study of cells. All cells have the same basic structure; they range in size from about 10 to 100 micrometres across, each one a bag of watery jelly held inside a very thin membrane, or wall, less than one-hundredth of a micrometre thick. The

---

* It has been suggested that Virchow plagiarised Remak, but it was certainly Virchow who popularised the idea, in his book *Cellular Pathology* in 1858.

cells that are most relevant to the story here, the ones that make up the structure of plants and animals, all have a central dark nucleus – physicists later borrowed the term to describe the central core of an atom. Although a single cell in isolation will form itself into a spherical shape, like a soap bubble, when they are joined to other cells they may be squeezed and stretched into other shapes. The cell wall ensures that each cell keeps its own identity, like bricks in a wall, but unlike bricks in a wall the membrane allows certain chemicals to pass in or out of the cell – in or out of each 'brick' – as required.

Focusing just on our kind of organism, the puzzle of life becomes attempting to understand how the fusion of one large cell, the egg, with one small cell, the sperm, can produce a single cell which then divides repeatedly in a complex process through a series of stages, which leads to the development of an adult being. By studying these stages of development through the microscope, biologists realised by the late nineteenth century that this development must be unfolding in accordance with some master plan – there was no miniature adult hidden inside the egg ready to be triggered into simple growth. But what was this master plan, and where was it hidden in the cell? This was the beginning of the road that led to the identification of DNA as 'the molecule of life'. The story really begins with experiments carried out by a Swiss biochemist, Friedrich Miescher (1844 to 1895), working at the University of Tübingen in the 1860s.

In 1866, Ernst Haeckel (1834 to 1919) had postulated that the factors that transmit heritable characteristics are contained in the nucleus of the cell. By that time, it was also known that proteins are the most important structural substances in the body – a fact reflected in their name, which means 'foremost'. Proteins are complex molecules with weights ranging from a few thousand to several million units on a scale where a single atom of carbon weighs twelve units, which gives you some idea of their size. They are made up from much smaller sub-units called amino acids, which themselves weigh in at, typically, a bit more than 100 units on the same scale. Just twenty different kinds of amino acids combine with one another, sometimes in large numbers, in complicated ways to make different proteins that are the fabric of life. These amino acids are themselves composed of atoms of carbon, hydrogen, oxygen and nitrogen (collectively known as CHON), and in one case sulphur.

Miescher wanted to identify the proteins involved in the chemistry of the cell, a key to the workings of life. The raw material that he used came from the pus-soaked bandages supplied by a nearby surgical clinic. He isolated the human white blood cells known as leucocytes from the pus, and found that the watery jelly that fills the cell is indeed rich in proteins. But then he found something new. When the cells were treated with a weak alkaline solution, his chemical tests revealed the presence of another substance, which was not a protein. By studying the cells under a microscope he found that the alkaline solution made the nucleus of a cell swell up and burst open; so the 'new' stuff he had found must be coming from the nuclei. The nuclei were not made of protein, but of a different material, which he dubbed 'nuclein'. Like protein, nuclein contained a lot of carbon, hydrogen, oxygen and nitrogen; but it also contained phosphorus, which is not found in any protein. Miescher wrote, 'I think that the given analyses – as incomplete as they might be – show that we are not working with some random mixture, but . . . with a chemical individual or a mixture of very closely related entities.' But he was not able to work out the structure of the large nuclein molecules. Miescher completed the first phase of this work in 1869, left Tübingen and wrote up his results for publication. But because of a chapter of accidents

– including the Franco-Prussian War – it was not published until 1871. In his later work, Miescher found that the nuclein molecules contain several acidic groups, and the term 'nucleic acid' began to be used to describe this material by the end of the 1880s.

By that time, there had been another significant advance in understanding the workings of the cell, partly stimulated by Miescher's work. Once the importance of cells as the basic units of life had been recognised, the key puzzle that needed attention was how individual cells divide and reproduce. Cytologists – people who study cells – used dyes to colour cells and highlight the structures within them. In 1879, the German biologist Walther Flemming (1843 to 1905) discovered that these dyes are taken up very strongly by thread-like structures inside the cell which become clearly visible during the process of cell division. Because of the ease with which these threads can be coloured, they became known as chromosomes, and other bits and pieces inside the cell were given names like chromatids and chromoplasts. By killing cells at different stages during the process of division, staining them with dyes and studying them under the microscope, Flemming found the pattern of events going on during the process, which he called mitosis. It took years to fill in all the details, but in essence what happens is that the chromosomes, which are usually packed inside the nucleus, are copied by the mechanism of the cell, with one set of chromosomes then going to one side of the cell, the other set to the other side and the cell splitting down the middle to produce two cells each with a complete set of chromosomes. There is no sense in which one cell can be identified as the 'daughter' and one as the 'parent'; each is an exact copy of the original. It was clear that chromosomes must be important to the cell, and it was soon realised that they must contain the blueprint, or instruction manual, for the workings of the cell. But it was also clear that this could not be the whole story – what happened when an egg and sperm cell fused to make the basis of a new individual? Why doesn't the fertilised egg have a double set of chromosomes?

The answer was provided, at least in outline form, by August Weismann (1834 to 1914), a zoologist based at Freiburg, in Germany, in the 1890s. In 1886, Weisman had suggested that the egg and sperm cells (together called the 'germ' cells, from the same root as germination) must contain some essential requirement for life that was passed on from one generation to the next. He followed this up by guessing (correctly) that this material of heredity must be carried in the chromosomes. He concluded that 'heredity is brought about by the transmission from one generation to another of a substance with a definite chemical and, above all, molecular constitution' which is found in chromosomes. And he saw that the only way to avoid hereditary material piling up in the cells of succeeding generations would be if the germ cells were produced by a special process of cell division, now called meiosis, which halves the quantity of hereditary material. The details were not worked out until later, but it makes sense to include them here. We now know that chromosomes come in pairs, associated with one another in the cell. In mitosis, each pair is copied and passed on as a unit. But in meiosis, they get separated. A slightly more complicated version of cell division occurs, first swapping some of the bits of material between the members of each pair of chromosomes* then producing two daughter cells, each with a full set of the newly shuffled chromosomes, before splitting again without any

---

* If you think of each pair of chromosomes as made up of a piece of red string and a piece of green string, in this process equivalent bits are chopped out of each strand and swapped, to produce two new bits of string, each with an alternating pattern of red and green along its length. Everywhere there is red on one string there is green on the other, and vice versa.

copying to make four cells each with a single set of unpaired chromosomes. When the egg and sperm fuse, the full set of chromosomes is restored, with the appropriate single strands from each germ cell getting together to make new pairs – crucially, half the chromosomes come from one parent, and half from the other parent. Specifically, in each *pair* of chromosomes, one member of the pair comes from one parent, and the other member of the pair from the other parent.

Except for the details of meiosis, this was the state of play at the time that the laws of heredity discovered by Mendel were rediscovered – not once, but by three different researchers, working independently of one another.

With the existence of chromosomes known, and their role in heredity suspected, it was natural for experimenters to turn to the kind of experiments Mendel had carried out, unknown to them, four decades earlier. As the nineteenth century came to a close, several researchers were independently carrying out this kind of work, and some of them even used peas in their studies, for the same reasons that Mendel chose those plants. The first of this new wave to publish was Hugo de Vries (1848 to 1935), working in the Netherlands with plants. In March 1900 he published two papers. The first one, in French, was a short summary of his results and made no mention of Mendel. The other, in German, offered a more detailed account, and did mention Mendel. He said of Mendel's paper 'this important monograph is so rarely quoted that I myself did not become acquainted with it until I had concluded most of my experiments, and had independently deduced the above propositions', although he does not say how he actually became acquainted with Mendel's work.[35] De Vries' French paper came as a bombshell to a German botanist, Carl Correns (1864 to 1933), who had been carrying out similar experiments, some of them with pea plants. Diligently checking the scientific literature before publishing his own results, he had found Mendel's paper; and then, before he could get his own work into print, de Vries beat him to the punch. An Austrian researcher, Erich Tschermak von Seysenegg (1871 to 1962), had a similar experience. In the end, it suited everyone to recognise Mendel as the pioneer of this kind of study, avoiding any unpleasant arguments about priority among the three of them. Confirmation of the significance of their (and Mendel's) work soon came from teams in the USA, England and France. By the end of 1900, Mendel's place in scientific history was established, as were his laws of heredity.

In the years following the discovery of the role of chromosomes in heredity and the rediscovery of Mendel's laws of heredity, nucleic acids were analysed and found to come in two varieties, familiar today, at least by name, even to non-scientists: DNA and RNA. Each type of molecule contains four sub-units, known as bases. In one of the acids these are called adenine, guanine, cytosine and thymine, often referred to by their initials as A, G, C and T. The other nucleic acid contains a different base, uracil (U), instead of thymine. But finding all this out took a long time. The 'discovery' of the life molecules RNA and DNA took a number of years and involved the work of many people. The person who actually gave these molecules their names was Phoebus Levene (1869 to 1940), a Russian-born American working at the Rockefeller Institute of Medical Research.

Levene began experimenting with nucleic acid, which he obtained from yeast cells, a few years after the breakthrough work of de Vries, Correns and Tschermak. This material contained about equal amounts of A, G, C and U, plus a chemical unit known as a phosphate group, essentially a phosphorus atom surrounded by four oxygen atoms. It also contained a carbohydrate group, a complex molecule made up of carbon, hydrogen and oxygen, but which carbohydrate had not been identified when Levene started work. In 1909 he isolated this substance and identified it as a sugar group, ribose. Sugar molecules are built around carbon rings, each with four carbon atoms and one oxygen atom linked to form pentagons. These units can then attach to other molecules to build up more complicated structures. Levene showed that the components of the nucleic acid are themselves linked in units, each made up of one phosphate, one sugar and one base, and he called these units nucleotides. But nobody knew how these components of the nucleic acids were joined together.

Levene had the idea that each nucleic acid molecule consists of a string of these nucleotides joined together, like the vertebrae of your spine. In 1909 he gave the name ribosenucleic acid to the molecule, which soon became known simply as RNA. Because there are four bases present in equal numbers in RNA, he guessed that each molecule is composed of a short chain of four units, called nucleotides, one associated with each of the four bases. In terms of the four bases, this would make many identical units spelling out something like A-C-U-G A-C-U-G A-C-U-G. This became known as 'the tetranucleotide hypothesis'. It turned out to be wrong, but it coloured thinking about nucleic acids for decades. In particular, it encouraged the idea that the really important molecules of life are all proteins; the nucleic acids were regarded as some sort of scaffolding to which protein molecules were attached.

It was another twenty years before Levene discovered, in 1929, that there is another kind of nucleic acid. Material derived from thymus cells turned out to contain a different sugar group, as well as having T instead of U. Because each molecule of this sugar group has one oxygen molecule less than a corresponding ribose group, he called it deoxyribose, and the nucleic acid became known as deoxyribosenucleic acid, or DNA. The names are often shortened slightly to ribonucleic acid and deoxyribonucleic acid. Levene still thought that the nucleotides in a DNA molecule must be linked in the same order, something like A-C-T-G, A-C-T-G, A-C-T-G instead of A-C-U-G and so on. But a year before he identified and named DNA the first clue that the nucleic acids provided more than mere scaffolding had already emerged; in order to pick up that thread, we need to backtrack a little once again.

A key step towards an understanding of how evolution works was taken by Thomas Hunt Morgan and his colleagues, working at Columbia University in the second decade of the twentieth century. Morgan was working with the fruit fly *Drosophila* rather than with peas, but in essence he was carrying out the same kind of experiments as Mendel. Pea plants only produce a new generation once a year, but the flies not only produce a new generation every two weeks, the females lay hundreds of eggs at a time, giving the researchers plenty of data to analyse. In these flies, the sex of an individual is determined by one of the chromosomes – which just happens to be very easily identified. There are two kinds of chromosome that between them determine the sex of the body, known as X and Y, from their shapes. In most species, the cells of females always carry the XX pair, while the cells of males carry the XY pair. Offspring must always inherit one X from the mother, and can inherit either X or Y from the father. If they inherit another X the individual will be female, if they inherit a Y they will be male. But Morgan discovered that this is not all that those chromosomes do.

Morgan started out with a population of flies that all had red eyes. But because of a chance mutation, in 1910 a single white-eyed male was noticed among thousands of flies being studied. Morgan mated the white-eyed male with a red-eyed female to see what would happen. All the offspring had red eyes. He then extended his studies into the grandchildren and succeeding generations, just as Mendel had done with peas. In the second generation, there were red-eyed females, red-eyed males and white-eyed males, but there were no white-eyed females. After carrying out a careful statistical analysis of his findings, in 1911 he concluded that whatever it was that was causing the white-eye mutation must be a factor carried on the X chromosome. In the second-generation females, even if one X chromosome has the mutation this is dominated by the normal factor on the other X chromosome; but in males there is no 'other' X chromosome to do the job. In further experiments the team showed that other properties of fruit flies are also linked with their sex, so must also be carried on the X chromosome. Morgan picked up the term 'gene', coined by the Danish botanist Wilhelm Johannsen (1857 to 1927) in 1905, for these Mendelian 'factors', and came up with an image of genes strung out along the thread-like chromosomes like beads along a wire.

The important point is that although each individual inherits one copy of each kind of gene from each parent, the two copies do not necessarily behave in exactly the same way. These different versions of a gene are called alleles. Getting back to Mendel's examples, but using modern terminology, there is a gene which determines colour, but it comes in two varieties; an allele for greenness (label it 'a') and an allele for yellowness (label it 'A'). One of the pairs of chromosomes inside the cells of the pea carries this colour gene, but the alleles on each of the two strands in the pair are not necessarily the same. The possible combinations are AA, Aa, aA and aa. Obviously, in AA peas the colour is yellow, and in aa peas the colour is green. But Aa and aA peas are not striped or spotted yellow and green, they are always yellow, because the A allele is dominant. Only instructions carried by the A allele are expressed, while the a allele is ignored. The same kind of behaviour occurs in many pairs of alleles.

Further work provided a picture of the way in which shuffling genes to make new combinations for sex cells occurs during meiosis. As we have mentioned, paired chromosomes are chopped up, with pieces being swapped from one chromosome to the other (termed 'crossing over'), then rejoined ('recombination'). The further genes are along the chromosome, the more likely they are to get separated when this process of crossing over and recombination occurs; genes that are close together tend to stay together. This made it possible, using a great deal of painstaking work, to map out the order of genes along the chromosomes for some species. But the key moment when the idea of Mendelian heredity and genetics became established was when Morgan and his colleagues published a classic book, *The Mechanism of Mendelian Heredity*, in 1915. Morgan continued his work on heredity, writing *The Theory of the Gene*, published in 1926, and receiving the Nobel Prize in Physiology or Medicine in 1933 'for his discoveries concerning the role played by the chromosome in heredity'. By then, the basis of the fusion of the law of natural selection with the laws of heredity had become established. This would later become known as the 'Modern Synthesis', although that name was not coined until 1942, by Julian Huxley (the grandson of 'Darwin's bulldog') with his book, *Evolution: The Modern Synthesis*.

Odd though it may seem to modern eyes, the rediscovery of Mendel's laws at the beginning of the twentieth century was initially seen as a blow to the Darwin–Wallace theory of natural selection. That theory was all about gradual change; but in their experiments people like de Vries saw sudden changes from one generation to the next – changes in colour, wrinkliness and so on. This, though, was because Mendel and the rediscoverers had deliberately sought out examples where there is a clear change visible from one generation to the next – yellow or green, wrinkly or smooth, and so on. Most characteristics in the majority of organisms are not inherited in this simple either/or way. People are not simply tall or short, but come in a variety of shapes and sizes, so-called phenotypes, built up from the interacting influences of a whole array of genes (the genotype). In order to investigate the effects of several alleles of the same gene at work on one characteristic of an organism, the Swedish geneticist Herman Nilsson-Ehle (1873 to 1949) studied varieties of wheat that could be cross-bred to produce kernels with any of five different colours, and found that the frequency of the occurrence of these colours exactly matched the statistical laws of Mendelian inheritance applied to the simultaneous transmission of *two* pairs of alleles located on two different pairs of chromosomes. Edward East (1879 to 1938), working at Harvard, carried out similar experiments on tobacco plants with short or long flowers.

All of this motivated the mathematicians to get in on the act. They realised that in a large population of individuals – such as the population of human beings on Earth – there may be a vast number of alleles of the same gene carried in different bodies. Each individual has no more than a pair of alleles for a particular characteristic; but there are many other versions of the same gene residing in the cells of other individuals. In principle, any of those alleles might be paired in the next generation. If something happened in the environment to make one particular allele advantageous, it would quickly spread through the population.

Today, for example, people are born with a variety of different eye colours, and there is no obvious evolutionary advantage in having, say, blue eyes. But if there were a change in the Sun's output which made blue eyes more efficient, so that blue-eyed people found it easier to find food (and we ignore modern technology), the allele for blue eyes would spread through the population and blue-eyed people would become more common. In the 1920s, four mathematicians were motivated to carry out the calculations that show how efficiently alleles can spread through a population. They were R. A. Fisher (1890 to 1962) and J. B. S. Haldane (1892 to 1964), each working in England, Sewall Wright (1889 to 1988) in the USA, and Sergei Chetverikov (1880 to 1959) in the USSR. The power of natural selection operating on a species in which the individuals between them carry a large number of different alleles was summed up in Fisher's book *The Genetical Theory of Natural Selection* in 1930. These studies showed that if a new allele, produced by mutation from an old one, gives the animals that possess it just a one per cent advantage over those that do not, then the new allele will spread through the entire population within a hundred generations. This is slow enough to match the geological evidence, but fast enough to explain the perfect camouflage of tiger beetles. An advantage which is too slight in individual terms even to be noticed by people studying a population of animals or plants in the wild is large enough to ensure the success of a mutated gene. Although the experts continued to debate the details, all intents and purposes the Modern Synthesis was established at the beginning of the 1930s, and the focus of our attention will now be on what goes on at the level of chromosomes, and the discovery of the role of DNA in evolution.

CHAPTER EIGHT

CRYSTALLISING THE ROLE OF DNA

At the same time that Thomas Hunt Morgan was getting to grips with the role of genes in heredity, experimenters in a seemingly unrelated field were developing the techniques that would eventually unveil the molecular mechanism of heredity. This is also a story of how a new scientific discovery can be quickly turned to experimental use, opening the way for more discoveries.

X-rays had been discovered in 1895, but at first their nature was something of a mystery. Nobody was sure if they consisted of a stream of particles, like electrons, or electromagnetic waves, like light but with a much shorter wavelength.*

---

* It later became clear that light and electrons each have both wave-like and particle-like properties, but that does not affect the story we tell here. For details, see John Gribbin, *In Search of Schrödinger's Cat*.

The breakthrough came in 1912, when a team headed by Max von Laue (1879 to 1960), working at the University of Munich, found that X-rays could be diffracted by crystals. When light is shone through two narrow slits in a screen, the waves spreading out on the other side of the screen make a pattern of light and shade, an interference pattern. Von Laue realised that the spacing of atoms in a crystal of zinc sulphide would make an array of 'slits' just the right size to produce the same sort of effect with X-rays. When his team carried out the experiment, they found a very complicated diffraction pattern, which was hard to interpret but was clear evidence of the wave nature of X-rays. The pattern produced on a photographic plate showed many distinct spots arranged in intersecting circles, centred on the spot produced by the main beam. Von Laue received the Nobel Prize in Physics in 1914 'for his discovery of the diffraction of X-rays by crystals'. But by then another team was well on the way to establishing the details of this diffraction process.

At that time, William Henry Bragg (1862 to 1942) was an established physicist working at Leeds University. His son, William Lawrence Bragg (1890 to 1971; he was always known as Lawrence), was just starting out as a research physicist in Cambridge. William had at first tried to explain the patterns found by the German team in terms of particles, but soon convinced himself that they had been produced by waves. Father and son discussed the implications with one another, and realised that it ought to be possible to work backwards from the diffraction pattern, analysing the arrangement of bright and dark spots to determine the structure of a crystal. Lawrence worked out the rules that determined where the bright and dark spots would be produced when a beam of X-rays with a particular wavelength struck a crystal made up of atoms spaced a certain distance apart from one another. This became known as Bragg's law. The law worked both ways. If you know the way atoms are spaced out in a crystal you can use diffraction to measure the wavelength of the X-rays. If you know the wavelength of the X-rays you can use diffraction to work out how the atoms in a crystal are arranged. Lawrence used his law to interpret the diffraction patterns obtained in Munich, but he did not have enough information about the wavelengths of the X-rays used in the experiment to make a detailed calculation. So William carried out more experiments, including the invention of the first X-ray spectrometer, an instrument that measured accurately the wavelengths involved. The data from these experiments could then be plugged in to Bragg's law. Once it was firmly established that X-rays behave as waves, they could be used to analyse the structure of crystals, which is where they later came into the story of DNA.

Interpreting the data is extremely difficult for complicated structures like DNA, in which there are large numbers of different kinds of atoms. But simpler crystals are easier to work with, and the technique soon showed, for example, that crystals of sodium chloride (common salt, NaCl) are made up not of lots of distinct NaCl molecules but of an array of sodium (Na) and chlorine (Cl) atoms with equal spacing, alternating in a lattice. The work of the two Braggs was described in a book, *X-rays and Crystal Structure*, published in 1915, while Lawrence was serving in the British Army in France. The same year he shared the Nobel Prize in Physics with his father 'for their services in the analysis of crystal structure by means of X-rays'. Lawrence was still only 25; he is the youngest person to receive the physics prize. He said in his Nobel Lecture:

> . . . the examination of crystal structure, with the aid of
> X-rays, has given us for the first time an insight into the
> actual arrangement of the atoms in solid bodies . . . There

*seems to be hardly any type of matter in the condition
of a true solid which we cannot attempt to analyse by
means of X-rays. For the first time the exact arrangement
of the atoms in solids has become known; we can see how
far the atoms are apart and how they are grouped.*

It was this ability that would in the decades that followed
lead to an understanding of the structure of proteins and
DNA. But that had to await the discovery of the central role
of DNA in heredity, which only began to become clear at
the end of the 1920s.

The next step involved experiments that once again provided
a more rapid timescale for changes to be studied. Mendel's
peas produced only one generation a year, limiting his opportunity to study heredity. Morgan's fruit flies reproduced every
couple of weeks. The next step, taken in 1928 by Frederick
Griffith (1879 to 1941), a medical officer working for the UK
Ministry of Health in London, involved bacteria, where scientists can see changes that take place in a matter of hours; it
also brought biologists a step closer to the key molecules
involved. Griffith was not primarily interested in genetics; he
was studying bacteria as agents of disease, rather than as a
tool for research into genetics. But along the way he made a
discovery that turned out to be crucial in understanding evolution.

The global influenza epidemic of 1918 to 1920 killed at
least fifty million people, more than the total battlefield casu-
alties of all the belligerents in World War I. In its aftermath,
governments around the world increased their research into
infectious diseases. Griffith's speciality was the study of pneumococci (a family of pneumonia-causing bacteria), with the
aim of developing a vaccine against pneumonia. In the early
1920s, he began working with two strains of pneumococcus,
which had very different effects on mice. In one strain the

bacteria are covered in a smooth coating (a polysaccharide), which makes cultures of the strain look shiny. This strain became known as 'smooth', or S. The other strain lacks this coating, and as a result cultures of the strain look rough and lumpy. The strain is called 'rough', or R. The S form is highly active and causes severe disease; but the R form is only weakly active and produces no more than a mild infection (there is a third strain of pneumococci, but these were not used by Griffith). Before Griffith's work, bacteriologists thought that each of the three strains of pneumococci were completely independent of each other, each fixed with its own properties down the generations. Griffith knew that different strains of pneumococci, some lethal and some not, could be present at the same time in the body of a person (or mouse) with pneumonia, and he carried out experiments to try to find out how this might affect the prospects of developing a vaccine.

When a body is infected with the rough strain of pneumococci, the bacteria are easily recognised as invaders by the body's immune system, and are killed off before any serious harm is done. The covering on the smooth strain seems to act as a camouflage that hides them from the immune system, so they can proliferate and cause serious illness, even death. Griffith showed that mice injected with the rough strain of pneumococci lived, while mice injected with the smooth strain died. He then injected mice with S bacteria that had been killed by heat treatment. The mice lived, but then came an astonishing result that he reported in January 1928.

In his next series of experiments, Griffith mixed harmless dead smooth bacteria with harmless live rough bacteria and injected them into mice. The mice died. Neither of the two forms alone was a killer, but the mixture was lethal. When he took samples from the dead mice he found that they were teeming with live smooth pneumococci. The live rough bacteria had been 'transformed', in Griffith's word, into live smooth bacteria. The explanation he proposed was that a transforming factor – what we would now call genetic material – from the dead smooth bacteria had been passed into the living rough bacteria. With the aid of this factor they had 'learned' how to develop a smooth coating. In further experiments, once the bacteria had been transformed they were transferred to a dish in the laboratory and monitored; the 'new' smooth bacteria reproduced to produce a colony of smooth bacteria, even though they were descended from transformed rough bacteria. As Griffith wrote in the scientific paper announcing the discovery, 'The R form . . . has been transformed into the S form'. But Griffith did not know which molecules were involved in the transformation. That only became clear after 1944, as a result of new experiments directly inspired by Griffith's observations.* By then, crystallography had already begun to reveal the structure of important biological molecules.

In the 1930s, it was still thought that proteins were the carriers of biological information, so these were the first biomolecules to be investigated by X-ray crystallography. Crystallography eventually showed that the key feature of these long chains of amino acids is the way in which they fold up into complex three-dimensional shapes, shapes that determine their biological properties.

The first steps towards this understanding were taken by J. D. Bernal (1901 to 1971) and his colleagues in Cambridge in 1934. Bernal, who had worked with William Bragg in the 1920s, started out using X-ray crystallography to determine the structures of graphite and bronze. When he tried to adapt these techniques to the study of organic molecules, he ran into a problem. The standard way of preparing crystals is to

---

* Griffith was killed in an air raid during the London Blitz of 1941 and did not live to see these developments.

grow them in a concentrated solution, known as the 'mother liquor'. Crystals are then allowed to settle out of this liquid as it evaporates – as in common schooldays' experiments involving things like common salt (sodium chloride) and copper sulphate. The individual molecules or atoms align themselves in a repeating series of 'unit cells' with a regular pattern, forming a crystalline 'lattice'. The researchers expected to be able to crystallise a protein in the same way, by allowing the purified protein to settle out of such a concentrated solution. But when proteins were dried out before being X-rayed, their structure collapsed, like a structured house of cards collapsing into a disordered heap.

John Philpot, a researcher from Oxford who was at that time based in Uppsala, in Sweden, was trying to crystallise the protein pepsin in the mid-1930s (pepsin is a digestive enzyme that breaks down other proteins in our food). He had prepared some crystals, growing in their mother liquor, and left them in the refrigerator in his lab when he went on a skiing holiday. When he got back he found that they had grown dramatically – some of them were 2 mm long. It just happened that at this time he was visited by Glen Millikan, from Cambridge, who reportedly took one look and said 'I know a man who would give his eyes for those crystals'. Philpot had plenty to spare, and generously gave Millikan some, still in a tube containing the mother liquor, to pass on to Bernal at the Cavendish Laboratory.

At that time, Bernal was collaborating with a visitor from Oxford, Dorothy Crowfoot (1910 to 1994; she later married and became known as Dorothy Crowfoot Hodgkin). Bernal found that when the crystals were fresh and damp they interacted with polarised light to produce a feature known as birefringence, which indicates an ordered crystal structure. So Bernal and Crowfoot sealed the crystal and its mother liquor inside a thin-walled glass tube (a capillary tube) and *then* studied it using X-rays. In this way, they obtained the first X-ray diffraction photograph of single pepsin crystals, in 1934. And Bernal's sealed capillary technique became the standard way to collect X-ray data on large biomolecules for the next fifty years.

It was clear from the beginning that photographs like this could in principle be interpreted to indicate the structure of the protein molecules themselves. When they described their experiment in the journal *Nature*, Bernal and Crowfoot wrote:

> *Now that a crystalline protein has been made to give X-ray photographs, it is clear that we have the means of checking them and, by examining the structure of all crystalline proteins, arriving at far more detailed conclusions about protein structure than previous physical or chemical methods have been able to give.*

Dorothy Hodgkin went on to develop the application of X-ray diffraction crystallography to the study of biologically important molecules over the next two decades, and received the Nobel Prize in Chemistry in 1964.* What eventually became clear, thanks to many studies by several people, is the complexity of the molecules of life. The order of the amino acids along a chain is only the primary structure of a protein. The chains can be twisted around to make a structure such as a helix, the secondary structure. And the helix or other secondary structure can be twisted into a kind of knot in three dimensions, the tertiary structure. The exact shape of the knot, not just its chemical composition, is what determines its role in the processes of life, but before the advent of high-speed computers, it was horrendously difficult and laborious

---

* Her life and work are superbly described in Georgina Ferry's biography of her.

to work all this out. Indeed, by 1971, only seven protein structures had been fully determined; but today the structures of more than 30,000 proteins have been identified. Back in 1944, however, the infant science of biomolecule crystallography was almost ready for its next challenge, when Frederick Griffith's 'transforming factor' was identified as DNA.

After Griffith's results were published in 1928, other researchers tried to find out just what it was that was being passed from one form of the bacteria to another. The key player in this work was Oswald Avery (1877 to 1955), who headed a team at the Rockefeller Institute in New York. Avery had been working on pneumonia since 1913, and was initially sceptical about Griffith's discovery, which seemed to fly in the face of the Rockefeller team's identification of distinct types of pneumococci. But their own experiments, and those of other teams, soon confirmed what Griffith had found, stimulating a new line of attack.

In 1931 the Rockefeller team found that the transformation process could occur even without involving mice. By growing R pneumococci in a Petrie dish (a standard shallow glass dish used in lab work), which also contained dead S pneumococci, they could transform the live R type into live S type. In order to identify the transforming agent, they first used alternate freezing and heating, to break apart the cells of a colony of S-type bacteria to make a liquid in which the interior contents of the cells were mixed with the outer fragments. The solid parts were settled out by spinning test tubes containing the mixture in a centrifuge, so that the solid pieces of cell fell to the bottom of the tubes, while the liquid containing the inner contents of the cells stayed above them. The liquid from inside the cells was enough to transform R pneumococci into the S type.

Establishing all this took time, but the details were clear by 1935. For the next stage of the investigation, Avery brought on board two other researchers, first the Canadian-born Colin MacLeod (1909 to 1972), then Maclyn McCarty (1911 to 2005), to work with him on a careful study of the genetically active liquid. The project took them nearly ten years to complete, as step by step they eliminated all the ingredients in the cell that were *not* causing the transformation, until they were left with only one possible culprit.

The first candidate for the transforming agent was protein. So the team attacked the liquid derived from S-type bacteria with a protease, an enzyme that chops protein molecules into little pieces. The liquid could still carry out the transforming process. They then looked at the possibility that the effect was linked to the smooth coating of the bacteria, compounds known as polysaccharides. They tested this using another enzyme that broke polysaccharides apart, but still found no effect on the transforming process. So they had to carry out a painstaking series of chemical processes to remove all traces of proteins and polysaccharides from the liquid before they began a detailed chemical analysis of what was left behind. The analysis showed that this had to be a nucleic acid, revealed by the proportions of carbon, hydrogen, nitrogen and phosphorus that it contained. The final set of experiments revealed that it was DNA, not RNA.

The discovery was published in 1944, leaving no doubt that the transforming agent was DNA. Avery's team did not actually say in print that DNA must be the material genes are made of, but Avery did speculate about this possibility privately, including in a letter to his brother Roy, a bacteriologist.[36] The suggestion that DNA, not protein, carried hereditary information was, however, so shocking that it was not immediately accepted by biologists at large. They were still largely convinced that DNA was too simple a molecule to do the job, and many of them thought that it was too big a jump from DNA as the transforming factor, revealed by

Griffith's work, to DNA as the active component in true genetics. It also took time for the news to spread, partly because of the disruption caused by World War II. But although biologists worldwide were dubious, the Avery–MacLeod–McCarty study stimulated more work by biochemists in the United States. It was the beginning of molecular genetics. Even so, it would be several years before the balance of evidence in favour of DNA as the genetic material became overwhelming, thanks to another brilliant experiment. Meanwhile, a new way of studying biomolecules was developed by another of the key participants in the story of DNA.

This new idea was the brainchild of the American chemist Linus Pauling (1901 to 1994), and occurred to him in 1948, when he was puzzling over the X-ray diffraction patterns produced by certain kinds of protein.

We have already mentioned one kind of protein, the globular molecules that are workers, such as the haemoglobin molecules that carry oxygen around in your blood. But there is another kind of protein, also based on long chains, known as polypeptides. In these fibrous proteins, the molecules are not folded up into a ball but largely retain the long, thin structure of a stretched-out chain. They are important as the structural material of the body, the basic components of things like hair, feathers, muscles, silk and horn.

The first X-ray diffraction images of fibrous protein were obtained in the 1930s by William Astbury (1898 to 1961),* working at the University of Leeds. He was studying keratin, a component of wool, hair and fingernails. The photographs did not provide enough detail to identify the exact structure of keratin, but they did show a regular repeating pattern, enough to indicate that the protein had a simple structure. In

---

* Astbury had studied under the supervision of William Bragg.

fact, he found two patterns; one, which Astbury called the alpha-form, corresponding to unstretched fibres, and another, dubbed the beta-form, which corresponded to stretched fibres.

Pauling had been the first person to work out the rules of quantum chemistry, and wrote a definitive book on the subject;[37] he was fascinated by the puzzle of using his understanding of chemistry to determine the structure of biomolecules, and later told how he 'spent the summer of 1937 in an effort to find a way of coiling a polypeptide chain in three dimensions, compatible with the X-ray data reported by Astbury'.[38] His first attempt to tackle the puzzle by looking at how the quantum chemistry of the atoms involved might link the components of the molecule failed, so he decided to go back to basics and study the structure of the amino acids that are the links in the chain before trying to work out how they fitted together. But this was not the only project he was working on in the 1940s, and like other researchers he was distracted by the disruption caused by World War II. It was a long time before the project came to fruition.

The first step was to study X-ray diffraction photographs of individual amino acids. Pauling carried this work out at Caltech (the California Institute of Technology) in collaboration with Robert Corey (1897 to 1971), and his understanding of quantum physics proved crucial. Many chemical bonds allow the atoms or chemical units on either side of the bond to rotate. But Pauling and Corey found that the peptide bond between carbon and nitrogen (which gives polypeptides their name) is locked by a quantum phenomenon known as resonance. A chain containing these bonds cannot rotate around them, so this part of the chain is held rigid.* This limits the number of ways in which the chain can be bent and folded. The

---

* Older readers may have come across a toy rather like this – Rubik's snake.

chain has two flexible links, then a rigid one, then two more flexible connections, then another rigid joint, and so on in a repeating pattern. But Pauling still could not work out how to fold up the chain to match Astbury's photographs, so he put the puzzle to one side until serendipity struck in 1948.

Although he was based at Caltech, that year Pauling visited the University of Oxford, in England. In the spring of 1948, he was in bed with a bad cold, and after he got bored reading science fiction and detective stories he amused himself by making another attempt to work out the structure of keratin.

With hardly any tools to work with, Pauling simply drew a representation of a long polypeptide chain on a long strip of paper. He remembered the distances between the various components, and the angles that the different units made with each other. But he found that it was impossible to make a chain built up in accordance with these parameters fit along the straight, flat piece of paper. One particular link, which occurred repeatedly at different places along the chain, always came out wrong. This link, which had to make an angle of 110 degrees, could not be changed because it was locked in place by the carbon-nitrogen quantum resonance. So if the link had to be fixed, the chain could not be straight. Seeing this with a flash of insight, Pauling creased the paper and folded it everywhere the crucial bond occurred to make the correct angle, 110 degrees. The creased strip of paper now made a corkscrew of repeating linkages that spiralled through space – a roughly helical shape. Even better, when he had got the angles just right the nitrogen-hydrogen group in one polypeptide bond fell in line with an oxygen atom attached to a carbon atom four steps along the chain. This happened all along the chain. Oxygen and hydrogen have an affinity, caused by quantum effects, which attracts them to one another through a so-called hydrogen bond. These hydrogen bonds would help to maintain the helical structure that Pauling had discovered.

Back in the USA, Pauling's team carried out further X-ray studies confirming that this single-stranded helix formed the basic structure of hair. His team then produced a tour de force – seven separate scientific papers, published in 1951, describing the structure of hair, feathers, muscles, silk, horns and other fibrous proteins in terms of what Pauling, borrowing Astbury's nomenclature, had dubbed the alpha-helix. But these details were less important than the way the break-through had been made. Pauling's success set people thinking about helices in the context of biological molecules, and it also highlighted the possibilities of the bottom-up approach, in which model-builders fitted the basic building blocks of biological material together to find a match with the X-ray data. Just two years later, this approach would snare the biggest prize in molecular biology, the structure of DNA.

Even at the end of the 1940s, in spite of the work of Oswald Avery and his colleagues, it was still widely thought that genetic information was carried by proteins, not by DNA. But then came the experiments that persuaded even the doubters that DNA is 'the' life molecule.

The scene was set by an analysis of DNA carried out by Erwin Chargaff (1905 to 2002), an Austrian-born researcher working at Columbia University in the USA. He was impressed by the Avery–MacLeod–McCarty work, and in the second half of the 1940s focused the attention of his laboratory on DNA. There are two kinds of bases that are incorporated into the structure of DNA and RNA. One kind, with a single, roughly hexagonal ring of six atoms that can attach to other atoms on the outside of the ring, are called pyrimidines. Uracil (U) and thymine (T) are members of this family. The other kind, the purines, have a more complicated structure, with two such hexagonal rings joined together along one side, in a rough figure-of-eight shape. Cytosine (C), adenine (A) and guanine (G) are members of this family. DNA contains only the bases C, A, G and T; RNA contains C, A, G and U. In a series of delicate experiments, Chargaff's team found that there is a set of simple rules which relates the amount of each base to the others in DNA. The rules were summed up in a paper published in 1950: the total amount of purine present in a sample of DNA (G + A) is always equal to the total amount of pyrimidine present in the sample (C + T); in addition, the amount of A is almost the same as the amount of T, and the amount of G is almost the same as the amount of C. The team also showed that the relative amounts of guanine, cytosine, adenine and thymine are different in different species.* This meant that DNA could not be a simple scaffolding with an endless repetition of the same four bases, but must have a more complicated structure; this was the death (not before time) of the tetranucleotide hypothesis. The 'Chargaff Ratios' would provide one of the keys to understanding the structure of DNA – but that understanding only came after another team proved beyond doubt that genetic information is carried by DNA.

The experimental steps along the road that led to an understanding of DNA used a succession of smaller and more rapidly reproducing organisms. Gregor Mendel worked with peas; Thomas Hunt Morgan worked with fruit flies; and Avery's team worked with bacteria. The final step was taken using the smallest entities that carry genetic material: viruses. The smaller an organism is, the less superstructure it carries and the more its genetic material dominates; viruses are the extreme of this progression.

Viruses are little more than bags of protein, far smaller than

---

* Human DNA is made up of 30.9% A, 29.4% T, 19.9% G, and 19.8% C.

a bacterium, filled with genetic material. They were first imaged in the 1940s, with the aid of electron microscopy; a typical virus has a structure rather like a tadpole, with a bag full of genetic material as the 'head', and a 'tail' which it uses to move around. When a virus attacks a cell, it makes a hole in the cell wall through which it squirts the genetic material into the cell, while the empty bag, or husk, remains attached to the cell wall. The injected material takes over the chemical factory of the cell and uses this to make copies of the virus out of the material inside it. Then the cell bursts open, releasing the copies of the virus to repeat the process.

This is life at its simplest – viruses exist only to make more viruses. Alfred Hershey (1908 to 1997) and Martha Chase (1927 to 2003), working at the Cold Spring Harbor Laboratory in the USA at the beginning of the 1950s, developed a neat experiment using viruses, which provided the ultimate proof that it was DNA that was carrying the instructions on how to make more copies of the virus into the cell being attacked.*

The viruses they worked with are known as bacteriophage (or phage, for short), because they 'eat' bacteria. The simple idea behind the experiment was based on the fact that phosphorus is present in DNA, but not in protein, while sulphur is present in protein but not in DNA. And both phosphorus and sulphur can be easily obtained (if you are a research scientist) in radioactive forms. Hershey and Chase 'fed' phage with bacteria that had themselves been allowed to reproduce in a medium which contained either radioactive isotopes of phosphorus (phosphorus-32) or those of sulphur (sulphur-35). The now-radioactive phage were then allowed to attack a colony of non-radioactive bacteria. The phages in the next

---

* Hershey received the Nobel Prize for this work in 1969; in one of the many examples of blatant sexism by the Nobel Committee, Chase was not included in the award.

generation, now laced with radioactive material, were used to infect non-radioactive bacteria, and these were then analysed. Everywhere the team detected phosphorus they would be tracing the path of DNA, while everywhere they detected sulphur they would be tracing the path of protein.

Unfortunately, after the radioactive phage had infected the culture of bacteria, the researchers were left with a mass of cells filled to bursting point with new viruses but with discarded phage husks – the bags that had contained the genetic material of the viruses – still attached to the walls of the bacterial cells. Both kinds of radioactive isotope were present in the brew. In order to separate out the leftover husks from the original generation of phage from the new viruses manufactured inside the bacteria, the team used an ordinary kitchen utensil known as a Waring Blender; to generations of biologists their work became known as 'the Waring Blender Experiment'.

They used the blender on a low setting to gently shake the phage husks loose from the cells they had infected. The mixture was then whirled around in a centrifuge, where the bacterial cells, full of new virus, fell to the bottom and could be extracted, while the husks of the old phage remained floating in the liquid left above them. Then the two components were analysed. Radioactive phosphorus, indicating the presence of DNA, was found in the cells (in the new generation of viruses), while radioactive sulphur, indicating the presence of protein, was found in the leftover husks. The results were published in 1952, leaving no more room for doubt. It is DNA that carries genetic information, and it is protein that is the building material of life.

The success of this superficially simple experiment owed a great deal to the expertise of Martha Chase, although officially she was 'only' the assistant to Alfred Hershey. Waclaw Szybalski, another Cold Spring Harbor biologist, later recalled:

Experimentally, she contributed very much. The laboratory of Alfred Hershey was very unusual. At that time there were just the two of them, and when you entered the laboratory there was absolute silence and just Al directing the experiments by pointing with his finger to Martha, always with a minimum of words. She was perfectly fitted to work with Hershey.[39]

It was now clear that the protein provided the structural material in the bacteriophage, while the DNA carried the genetic information. After this, hardly any biologists believed that the genetic material could be anything but DNA, and the stage was set for the revelation of the structure of DNA itself.

Even while Hershey and Chase were carrying out their experiments in the USA, researchers in England were closing in on the structure of DNA. The basic structure was first revealed by experiments carried out by a team working at the Medical Research Council's Biophysics Research Unit at King's College London, although because of a curious twist of fate they did not receive appropriate credit for their work at the time. The head of the unit, John Randall (1905 to 1984), was one of the first people to accept the evidence that genetic information is carried by DNA; he was a physicist who had been trained by Lawrence Bragg and knew X-ray diffraction, but his unit was a pioneering group, unusual in those days, where biologists, biochemists and scientists from other disciplines worked with physicists.

In 1950 New Zealand-born Maurice Wilkins (1916 to 2004) was working in the unit, studying various different kinds of biological molecules, including DNA, proteins, tobacco mosaic virus and vitamin B12. In May that year, the Swiss biochemist Rudolf Signer (1903 to 1990) came to a meeting of the Faraday Society in London to report on his success, with the help of his student, Hans Schwander, in the extraction of nucleic acids from the thymus glands of calves. He gave Wilkins a sample of very pure DNA. This was not exactly a bolt from the blue – Signer had been working with DNA for years, and back in 1938 he had reported, in a paper published in *Nature*, that what he then called thymonucleic acid must be a long, thread-like molecule, weighing in at 500,000 to 1,000,000 units on the usual scale. But like so much 'pure' scientific research, developments and the spread of information were delayed by World War II. The DNA Wilkins had to work with was in the form of a gel, and he was preparing some for analysis using ultraviolet light when he noticed, as he put it in his Nobel Lecture:

> [that] each time that I touched the gel with a glass rod and removed the rod, a thin and almost invisible fibre of DNA was drawn out like a filament of spider's web. The perfection and uniformity of the fibres suggested that the molecules in them were regularly arranged. I immediately thought the fibres might be excellent objects to study by X-ray diffraction analysis. I took them to Raymond Gosling, who had our only X-ray equipment (made from war-surplus radiography parts) and who was using it to obtain diffraction photographs from heads of ram spermatozoa.

Gosling (1926 to 2015), who was a PhD student at the time, worked under the direction of Randall. He kept the DNA fibres moist (remembering the work of Bernal on proteins) and sealed them in capillary tubes filled with hydrogen so that there would be no interference in the X-ray patterns from atoms such as carbon and nitrogen found in the molecules that make up air. Although Gosling did obtain diffraction images of DNA in 1950, there were limits to what

could be done with the cobbled-together equipment he had.*
But things began to change, in more ways than one, in 1951.
Randall not only obtained new equipment, but recruited
another researcher, Rosalind Franklin (1920 to 1958), to tackle
the problem of determining the structure of DNA.

Franklin was an expert at X-ray crystallography, and had
been working in Paris on the structure of coal and compounds
derived from coal, but had not previously tackled biological
molecules. She was recruited on a three-year appointment to
work on proteins and lipids, but instead, when she arrived at
King's in January 1951, Randall had already decided to give
her the job of analysing the Signer DNA, with Gosling as
her assistant. Wilkins, understandably, was rather put out by
this arrangement, and when he protested Randall allowed him
to carry out his own studies using a sample of DNA provided
by Chargaff. Also in 1951, in May, Wilkins presented some
of the images he had obtained with Gosling at a meeting in
Naples, where they fired the interest of one of the junior
participants – the American James Watson (born 1928).
Watson had recently completed his PhD and was spending a
year working in Copenhagen, but would soon move to
Cambridge.

Franklin and Gosling combined well together. Gosling was
able to crystallise DNA – the first person to crystallise genes
– and Franklin was able to tweak the new equipment to
produce peak performance. They obtained the first X-ray
diffraction images of crystalline DNA, and discovered that
there are two forms. When it is wet, it forms a long, thin

---

* Astbury had obtained diffraction images of DNA back in 1938, showing
that it had a regular structure. The results were not detailed enough to
determine that structure, although in a paper published with Florence Bell
(1913 to 2000) they described it as like a 'pile of pennies'. Their work was
interrupted by World War II, when Bell served as a radio operator in the
Women's Auxiliary Air Force (WAAF), and was never followed up.

fibre, but when it is dry, it is short and fat. These became known as the 'B type' and 'A type', respectively. Because of the humidity inside cells, the B type was expected to be more like the DNA found in living things.

Because of the rivalry between the researchers at King's, at Randall's direction Franklin focused on the A type, while Wilkins concentrated on the B type. The data eventually showed evidence of a helical structure for both types, and Franklin gave a talk summarising the work carried out so far at King's in November 1951. Her lecture notes include the statement:

> *The results suggest a helical structure (which must be very closely packed) containing 2, 3 or 4 co-axial nucleic acid chains per helical unit . . .*[40]

Watson was present at this talk, and also discussed DNA with Wilkins – at that time, the helical structure was more firmly established for the B type of DNA than for the A type. But he has always claimed that he had no recollection of this comment by Franklin. Alex Stokes (1919 to 2003), a colleague at King's, actually suggested a double-helix structure in which the bases sticking out from the sugar-phosphate spines were stacked, using the same analogy as Florence Bell, like 'a pile of pennies', or like playing cards being riffled together in a shuffle. But the details were far from clear, and determining the exact structure required a lot more data, and a lot more analysis, which took up most of 1952.

At the beginning of 1953, when it was clear that both forms of DNA were based on a helical structure, Franklin prepared two scientific papers suggesting a double-helix structure for the A type. These were submitted to the journal *Acta Crystallographica*, where they arrived on 6 March that year. This was Franklin's swan song at King's, before she moved on to Birkbeck College, also in London. She also drafted another paper, dated 17 March 1953, which presented the evidence for a double-helix structure for B type DNA. But this paper was never published in its original form, and only became known after her death, because it was overtaken by events in Cambridge. Just at that time, the structure of the B type DNA had also been determined by another team, who in a breach of professional etiquette had been given access by Wilkins to Franklin's data, unknown to her; this included a print of one of her (and Gosling's) best diffraction images. The image, obtained by Gosling in May 1952 and known as Photo 51, showed the X-ray diffraction pattern of DNA in the highest quality available at the time. Its clearest feature is a cross-shaped pattern that could only be produced by a helical structure. If one image could be said to have unlocked the secret of the structure of DNA, it was this one.

The person who was shown that image, in January 1953, and hurried back to Cambridge with it, was James Watson. In his book *The Double Helix*, Watson says, 'The instant I saw the picture my mouth fell open and my pulse began to race. The pattern was unbelievably simpler than those obtained previously . . . the black cross of reflections which dominated the picture could only arise from a helical structure'. The reason for his excitement was that in Cambridge Watson had teamed up with Francis Crick (1916 to 2004), to attempt to determine the structure of DNA. Crick was a physicist who had become disillusioned with the subject by his war work and the development of the nuclear bomb, and shifted into biology, joining the Medical Research Council Unit at the Cavendish Laboratory in 1949, at the age of thirty-three, to work for a PhD. Although he was awarded the degree in 1953 for a study of polypeptides and proteins, this was just after he achieved fame, alongside Watson, for his unofficial work on the structure of DNA.

Watson and Crick had been allocated space in the same room, where Crick was soon fired up by Watson's enthusiasm and began moonlighting on the puzzle, in collaboration with Watson. Compared with the people at King's, they were both amateurs – dilettantes who knew little about the history of DNA studies but were strongly influenced by Pauling's bottom-up approach. In between their proper studies (Watson was supposed to be working on the tobacco mosaic virus) they tried building models of the molecule, but were hamstrung by their ignorance. A breakthrough came in June 1952, after Crick discussed the problem with John Griffith (1928 to 1972), the biochemist nephew of Frederick Griffith. Crick had picked up on the idea that the flat bases from opposite strands of DNA might stack up above one another like a riffled pack of cards, and he asked Griffith if he could work out which of the bases might fit together in such a stack. In the tea queue at the Cavendish one afternoon a few days later, Griffith said that he had looked at the chemistry and found that adenine would naturally link up with thymine and guanine with cytosine. Crick immediately realised that this would allow what is called complementary replication – if the strands are pulled apart, breaking up the CT pairs and the AG pairs, then everywhere there is a C on one strand it will link up with a loose T, everywhere there is a G it will link up with A, and so on. Griffith also had the same insight, but with his expertise in chemistry he had already realised something that Crick did not immediately appreciate. The properties Griffith had found did not encourage the bases to stack one on top of the other. Both CT pairs and AG pairs could link *edge to edge* using a combination of hydrogen bonds, and each of these pairs had the same width, so they would take up the same space when joining DNA strands together, like the rungs on a helical ladder.

It's a measure of just how ill-informed Crick and Watson

were at this time that they had still not heard of the Chargaff Ratios.* But in July 1952 Chargaff himself visited Cambridge, where Crick asked him if anything really useful had come out of all the chemical analysis of DNA. We pick up the story from an interview Crick gave to Robert Olby in 1968:

> *Chargaff, slightly on the defensive, [said] 'Well of course there is the 1:1 ratios.' So I said: 'What is that?' So he said: 'Well it is all published!' Of course I had never read the literature, so I would not know. Then he told me, and the effect was electric. That is why I remember it. I suddenly thought: 'Why, my God, if you have complementary pairing, you are bound to get a one to one ratio.' By this stage I had forgotten what Griffith had told me. I did not remember the names of the bases. Then I went to see Griffith and I asked him which his bases were and wrote them down. Then I had forgotten what Chargaff had told me, so I had to go back and look at the literature. And to my astonishment the pairs that Griffith said were the pairs that Chargaff said.*

Armed with this information, Photo 51 and other data from King's, early in 1953 Crick and Watson were led to their famous model of DNA in which everything fits together if each molecule consists of two strands twined around each other in a double helix, with the bases on the inside, so that the bases on one strand link up with the bases on the other strand. Adenine always links with thymine, and cytosine always links with guanine. Because the strands are like mirror images of each other, if they unravel each lone strand can

---

* At least, Crick did not remember having heard of them. Watson later claimed that he had mentioned them to Crick. If so, at first neither of them realised their importance at the time.

build a new double helix by adding the appropriate units to build up the partner strand.

The structure also carries information. The A, T, C and G can occur along the strand in any order, such as AATCAGTCAGGCATT . . ., like a message in a four-letter alphabet. Even binary computer code, a simple two-letter 'alphabet', can carry a great deal of information (including all the information in this book), and four letters are ample to contain all the information of heredity, provided the messages are long enough. Crick and Watson completed their model building on 7 March 1953, and sent a paper off to *Nature* – the day *after* Franklin's two papers arrived at *Acta Crystallographa*. When a draft of the paper reached Wilkins, after Franklin had already left for Birkbeck, he proposed that his team might publish a short item alongside the Crick and Watson paper 'showing the general helical case',[41] and casually mentioned that since Franklin and Gosling had also come up with something they had planned to publish, there ought to be 'at least 3 short articles in *Nature*'.

The '3 short articles' appeared in the 25 April issue of the journal. First came the Crick and Watson paper, claiming that the model was inspired by Chargaff's Ratios and mentioning the X-ray data as supporting evidence, rather than acknowledging that the model was actually inspired by the X-ray data and confirmed by Chargaff's Ratios.* Then came a paper by Wilkins, Stokes and their colleague Herbert Wilson (1929 to 2008) presenting the general case for X-ray studies as supporting the overall idea of a helical structure. Bringing up the rear was a paper by Franklin and Gosling which included the key Photo 51 that was so important to the discovery of the Watson–Crick model. Nobody, certainly not Franklin,

---

* They did have the decency to include a footnote noting that they had 'been stimulated by a general knowledge of' Franklin and Wilkins' work.

could have guessed from the presentation of these papers that Photo 51 had played an important part in the Watson–Crick model building. The third paper was, in fact, a slightly adapted version of the one completed on 17 March, the day before Wilkins had written to Cambridge with his '3 short articles' proposal. It gives details of the specific double-helix structure, but without the inspired idea of the base pairing mechanism.

In 1962, Crick, Watson and Wilkins shared the Nobel Prize in Physiology or Medicine for this work. It has sometimes been suggested that Franklin was another victim of anti-female prejudice by the Nobel Committee; but she had died of cancer, possibly caused by her work with X-rays, in 1958. She could not share the honour, even if the Committee had wanted to recognise her, as Nobels are never given posthumously. If anyone deserves our sympathy in this regard, though, it is surely Gosling, whose work in crystallising DNA and obtaining the diffraction images was of key importance. Nobel Prizes have certainly been awarded for less.

Even the identification of DNA as the carrier of the genetic code was not, however, the end of the story of the development of an understanding of evolution. Biologists still had to crack the code and find out how genetic information is passed from the DNA in chromosomes to the workings of the cells. This led – and is still leading – to new insights into evolution, including some surprises. It turns out that while Lamarck certainly was not entirely right, he may not have been entirely wrong, either.

CHAPTER NINE

THE NEW
LAMARCKISM

Once it became clear that DNA carries the code of life – the set of instructions that are used by the cell to make the proteins that actually do the work of life, as well as forming the structure of an organism – there was an intense effort to 'crack the code' to find out how these mechanisms work. This took many years, and involved many teams of researchers carrying out cunning biochemical experiments which we do not have space to describe in detail here. But we can at least explain the principles behind this work, and the results of all that effort.

The story of DNA code-breaking really begins with a book by a physicist, not a biologist. The quantum pioneer Erwin Schrödinger (1887 to 1961) became fascinated by the idea that quantum processes could be important in inducing changes in the molecules that carried the code of life

– mutations. At that time, in the 1940s, it was still widely thought that proteins carried the genetic code, but Schrödinger's ideas, published in 1944, did not depend on exactly which molecules were involved. He drew a distinction between a crystal of a substance such as common salt, where there is an endless repetition of the same pattern of sodium and chlorine atoms, and what he termed an aperiodic crystal, with a structure like 'say, a Raphael tapestry, which shows no dull repetition, but an elaborate, coherent, meaningful design', even though that design is made from threads with a limited number of colours. He referred to the information carried by the molecules of life as a 'code-script', and pointed out how even a limited number of letters in such a code-script (individual molecular groups, for example) could spell out information as efficiently as the individual letters of an alphabet. He said, 'the number of [different] atoms in such a structure need not be very large to produce an almost unlimited number of possible arrangements', and pointed out that in the Morse code two different signs (dot and dash) used in groups of four allowed for thirty different code groups – enough to cover the English alphabet and some punctuation symbols. Jumping ahead of our story slightly, four different signs in various combinations can be written in twenty-four different ways ($4 \times 3 \times 2 \times 1$), and twenty different groups can be arranged in approximately 24×10^{17} (24 followed by 17 zeroes) different ways. A four-letter code is sufficient to specify each of the twenty amino acids used in proteins; twenty different amino acids are ample to account for the variety of proteins in living things.

Schrödinger's book, *What is Life*, had a big influence both on biologists and on physicists who had seen enough of death in World War II and wanted to work on life. Among the people who later specifically recalled that it had influenced them were Maurice Wilkins, Erwin Chargaff, Francis Crick and James Watson. And just after Watson and Crick published their first papers on DNA, another physicist, George Gamow (1904 to 1968), got in on the act.

It was actually a second paper on DNA written by the Cambridge team, published in *Nature* on 30 May 1953,[42] that caught Gamow's attention. At that time he was visiting the Berkeley Campus of the University of California, from his base in Washington. He later recalled:

> I was walking through the corridor in Radiation Lab, and there was Luis Alvarez going with *Nature* in his hand . . . he said, 'Look, what a wonderful article Watson and Crick have written.' This was the first time that I saw it. And then I returned to Washington and started thinking about it.*

The fruits of that thinking appeared in *Nature* in February 1954. Gamow latched on to the discovery that the makeup of DNA involves four different kinds of base strung out aperiodically along a fibre, and highlighted the idea that protein molecules could be built up from chains of amino acids held in place along a strand of DNA, with each amino acid lining up next to a particular DNA-base code group. The details of his proposed mechanism were wrong. But as he spelled out:

> . . . the hereditary properties of any given organism could be characterised by a long number written in a four-digital system. On the other hand, the enzymes, the composition of which must be completely determined by the deoxyribonucleic acid molecule, are long peptide chains formed by about twenty different kinds of amino

acids, and can be considered as long 'words' based on a 20-letter alphabet.

Two key facts eventually emerged from all the painstaking work that followed this realisation. First, amino acid chains are not built directly on DNA. When the cell needs a particular protein (and how it 'knows' when it needs it is still largely a mystery), just the relevant bit of DNA is untwisted from the double helix of one chromosome and used as a template on which to build a strand of RNA, before being coiled up and packed away again. That strand of RNA is then used as the template to make a protein, before being disassembled ready for its parts to be re-used. Second, although there are four letters in the genetic code, they are actually used to make three-letter words, each of which specifies a particular amino acid, or in some cases the instruction to 'start' or 'stop' building a new chain. Because it is RNA, not DNA, that is directly involved in making proteins, the four letters involved are U, C, A and G. For example, the triplet AGU codes for the amino acid Serine, GUU corresponds to Valine, CCA means Proline, and UAG means 'stop'. So a string of bases along an RNA molecule, such as UCCAGUAGCGGACAG, should actually be read as UCC AGU AGC GGA CAG.

To see how this affects evolution, we can use a similar example from our own familiar alphabet. A message in three-letter words might read THE BAT HAS ONE HAT THE CAT HAS TWO. A simple mutation, changing just one letter, might make a nonsense word in the chain – THT BAT HAS ONE HAT THE CAT HAS TWO – which might or might not be important to the workings of the cell. Or it might make a different meaningful word ('meaningful' in the sense that it codes for another amino acid) – THE CAT HAS ONE HAT THE CAT HAS TWO. That altered amino acid might result in the manufacture of a useless protein. Or it might, just occasionally, cause the manufacture of a protein that works more effectively than the original. More extreme 'mutations' might change whole 'words' – HAT becomes ONE, perhaps; or delete words altogether – THE BAT ONE HAT THE CAT HAS TWO. And omitting (or adding) a single letter changes the entire message. Take out the first 'E', for example, and we are left with THB ATH ASO NEH ATH ECA THA STW O.

We will leave you to play with other examples. What matters for evolution is that mistakes of this kind can be produced as copying errors when chromosomes are being chopped up and rearranged before being separated into the germ cells which carry the genetic code into the next generation. There may also be more dramatic changes, for example, when bits of DNA get put back the wrong way round after crossing over, or get left out altogether. But we do not need to go into those details. What matters here is that the source of the not-quite-perfect copying of genetic material that is the basis of evolution has been discovered. With that in mind, we can look again at the evolutionary behaviour of whole organisms, and the insights that emerged in the second half of the twentieth century.

The foundations for those insights were actually laid down, but not widely appreciated at the time, in the 1930s. Although by that time the main focus of attention was on ever smaller units of life, one person in particular continued to focus her attention on much larger organisms, in the tradition of Gregor Mendel. She was Barbara McClintock (1902 to 1992). The organisms she worked with were not peas but maize plants, but like Mendel's peas they produced just one generation each year. She was also like Mendel in another way – her work produced insights that were not fully appreciated for four decades, although unlike Mendel she lived to see her work come in from the cold.

McClintock was born just two years after the rediscovery of Mendel's laws, and studied at the New York State College of Agriculture and Life Sciences on the campus of Cornell University in Ithaca, New York, graduating in 1923. As she put it in her Nobel Lecture in 1983, 'I became actively involved in the subject of genetics only twenty-one years after the rediscovery, in 1900, of Mendel's principles of heredity, and at a stage when acceptance of these principles was not general among biologists'. She did postgraduate work at Cornell, received her PhD in 1927, developed techniques for analysing the chromosomes of maize, and continued this line of research after earning her doctorate. It didn't matter for these investigations what the chromosomes are made of, because McClintock and the team she built up were interested in whole chromosomes and genes – sections of chromosomes – and their influence on the organism they inhabited. The organism she chose, maize, is much more interesting than the bland appearance of the uniform yellow heads of sweetcorn on the supermarket shelves might suggest. Wild maize produces seeds in a variety of colours, and the seeds are displayed in full view in rows along the kernel. Instead of having to catch tiny flies and look at their eyes, or study bacteria under the microscope, all you have to do to identify changes (mutations) is peel back the husk and see the patterns of coloured seeds displayed in full glory. But it was still necessary to use microscopes to study the genes themselves. McClintock developed improved techniques for staining maize chromosomes to make them visible, and used these techniques to reveal the morphology of the ten chromosomes found in maize. The most significant discovery of this early phase of McClintock's research was made in 1929, with the help of a research student, Harriet Creighton (1909 to 2004). In one strain of corn that they studied, the corn kernel could be either dark or light in colour, corresponding to the presence of a chromosome that came in two alleles that differed slightly from one another (such a pair is said to be heterozygous). This kind of behaviour had been inferred before, notably in Thomas Hunt Morgan's work on fruit flies. But the existence of the different alleles had indeed only been *inferred* from those studies. McClintock and Creighton went one better; by staining chromosomes and studying them under the microscope they found that the difference between the two kinds of maize showed up as a visible difference between the alleles. The relevant chromosome in the dark plants had a 'knob' that was missing on the equivalent chromosome in pale plants. This was the first direct observational evidence that differences in chromosomes affected the whole organism – the phenotype. When Morgan visited Cornell and learned of this work, which formed the basis of Creighton's PhD thesis, he urged them to publish it more widely, and it appeared in the Proceedings of the National Academy of Sciences in 1931. Just two years later, Morgan received the Nobel Prize 'for his discoveries concerning the role played by the chromosome in heredity'.*

Among McClintock's other early achievements, she showed how specific groups of chromosomes work together to produce traits that are inherited together, studied the way recombination directly observed under the microscope correlates with new traits, and after working at Missouri with geneticist Lewis Stadler (1896 to 1954) in the summers of 1931 and 1932, she used X-rays to increase the rate of mutation in maize and study the results. After a series of short-term appointments, including a brief spell in Germany, at the end of 1941 she was offered a permanent post in the Carnegie Institution's Department of Genetics at the Cold Spring Harbor Laboratory. Impressive

---

* From a modern perspective, it is surprising that McClintock, at least, did not share the award. She would go on to win her own Nobel Prize in 1983 for the discovery of genetic transposition.

though her earlier achievements were, it was there that she carried out her most important work.

The key discovery stemmed from a simple observation – in one line of plants the leaves were not always uniform in colour, but might have patches marked with a different colour. In most maize plants the leaves are green, but in some strains the leaves may be pale yellow, in others a lighter shade of green, or even white. But in some individual leaves there may, for example, be a streak of darker green on a light green leaf, or a patch of yellow on a green leaf. This intrigued McClintock because she knew that each leaf develops from a single cell at the stem of the plant. The leaf grows by the repeated division and multiplication of cells from this single source. So the patches of odd colour could be traced to a copying error – or mutation – in the chromosomes of a single cell, which then produced daughter cells containing a slightly different set of genetic instructions, copied faithfully in subsequent generations to make the streak of 'wrong' colour. McClintock could identify exactly which cell was involved in the mutation, and exactly when during the process of development and differentiation the mutation had occurred.

That wasn't all. Some of these multi-coloured leaves had a different pattern of mutational changes than others. The rate of mutations could be faster or slower depending on which leaf was involved. This could also be traced to a change in the genetic code carried by the chromosomes of a single cell at an early stage of the process of leaf differentiation. And similar effects were seen in the seeds of the corn cobs, involving the frequency and position of various colours among the seeds.

By 1947, after years of study similar to the work of Mendel but aided by direct observation of chromosomes under the microscope, McClintock had an explanation for what is going on. The genes that are responsible for the structure and working of an organism are not 'on' all the time (leaves do not keep growing forever, for example), but are copied into RNA, and then protein, only when required. This means that they must be controlled by other genes, responsible for turning them on and off. Without knowing about the roles of RNA and DNA, the existence of control genes was becoming clear in the 1940s. McClintock realised that there must be two kinds of control genes. One sits next to a structural gene, on the same chromosome as the gene it controls, and turns it on or off (or from green to yellow, of course). But her studies showed that there must be another kind of gene (or controlling element, as McClintock called them) that determines the rate at which the first control gene operates, speeding up or slowing down the frequency of changes in the system it controls. Her research showed that although the first kind of control gene sits on the chromosome next to the gene it controls, the second control gene (the regulator) can be almost anywhere in the cell – far away on the same chromosome, or even on a different chromosome altogether. In her further work up to the end of the 1940s McClintock showed that these regulators didn't even have to stay on the same chromosome. They appeared to be able to jump from one place on a chromosome to another, or even from one chromosome to another inside a cell, bringing different structural genes and other controllers under their influence. It is now clear that these regulators do not literally jump from one place to another, but that they can be copied by the mechanism of the cell and the copies inserted into different places on the same or different chromosomes. But the term 'jumping genes' has become a common shorthand to describe the process. The key point is that even within a single cell, the genome may not be fixed and unchanging. McClintock's work also indicated how cells with identical genomes can have different functions in an organism. It's worth quoting again from her Nobel Lecture in 1983:

> It soon became apparent that modified patterns of gene expression were being produced, and that these were confined to sharply defined sectors in a leaf. Thus, the modified expression appeared to relate to an event that had occurred in the ancestor cell that gave rise to the sector. It was this event that was responsible for altering the pattern and/or type of gene expression in descendant cells, often many cell generations removed from the event. It was soon evident that the event was related to some cell component that had been unequally segregated at a mitosis. Twin sectors appeared in which the patterns of gene expression in the two side-by-side sectors were reciprocals of each other.
>
> For example, one sector might have a reduced number of uniformly distributed fine green streaks in a white background in comparison with the number and distribution of such streaks initially appearing in the seedling and showing elsewhere on the same leaf. The twin, on the other hand, had a much increased number of such streaks. Because these twin sectors were side-by-side they were assumed to have arisen from daughter cells following a mitosis in which each daughter had been modified in a manner that would differentially regulate the pattern of gene expression in their progeny cells. After observing many such twin sectors, I concluded that regulation of pattern of gene expression in these instances was associated with an event occurring at a mitosis in which one daughter cell had gained something that the other daughter cell had lost.

> *Believing that I was viewing a basic genetic phenom-*
> *enon, all attention was given, thereafter, to determine*
> *just what it was that one cell had gained that the other*
> *cell had lost.*

By the beginning of the 1950s, McClintock was a senior and respected scientist. But when she published a paper describing her work in the *Proceedings of the National Academy of Sciences* in 1950, and followed this by presenting these discoveries at an annual meeting known as a Cold Spring Harbor Symposium, in the summer of 1951, they fell flat. The role of the gene was not yet fully understood, and nor was the role of DNA, so McClintock's talk of controlling genes and regulators did not involve a language her peers understood. In a sense, she was ahead of her time; but paradoxically she was seen as being behind the times – a kind of latter-day Mendel dealing with plants, who was not part of the mainstream of people involved in the brave new world of studying evolution using bacteria, viruses and X-ray crystallography. The result was that her work was essentially ignored. In return, McClintock essentially ignored everyone else, and continued to plough her own furrow. Among other things, she identified the existence of 'suppressor' genes, controlling elements that inhibit the activity of some functional genes. After 1953, however, she stopped publishing her results, except in the annual reports of the work of the Cold Spring Harbor Laboratory. In another echo of Mendel's story, the significance of her work was only widely appreciated after someone else made essentially the same discovery independently; but at least this rediscovery happened in her lifetime.

The key work was carried out by the French pair Jacques Monod (1910 to 1976) and François Jacob (1920 to 2013), using *E. coli*. From studies of mutant strains of these bacteria, at the beginning of the 1960s they discovered the same pattern of behaviour of control genes that McClintock had discovered a decade earlier from her studies of maize. When they published their results, in 1961, their paper in the *Journal of Molecular Biology* made no mention of McClintock, because they were unaware of her work; but others soon pointed out the connection, and as studies of the way control genes operate took off, opened up by a growing understanding of the relative roles of DNA and RNA, the importance of her contribution was increasingly appreciated. Even so, it was not until 1983, when she was eighty-one years old, that McClintock was awarded her Nobel Prize, eighteen years after Jacob and Monod had shared the award with André Lwoff (1902 to 1994), another French microbiologist who did ground-breaking work on bacteriophages. It is the understanding of the way in which bits of DNA can be copied from one chromosome and spliced into another that has made it possible to carry out genetic engineering, using the cells' own mechanisms to replace faulty genes in people suffering from certain illnesses, and to develop improved crops.

By the 1980s it was clear that the genomes of complex organisms such as ourselves, or oak trees, are not fixed, but are in a state of dynamic change. Genes are rearranged among the chromosomes as a matter of routine, on an evolutionary timescale, and this is one of the driving forces of evolution, providing some of the variety on which natural selection acts. This has led to two new insights into the nature of evolution, and neither of these in any way diminishes or undermines the work of Darwin and Wallace. Natural selection operates on varieties exactly in the way they discovered. But neither Darwin nor Wallace (nor anyone else in the nineteenth century) knew exactly how the variety on which selection operates is produced, and that is where the new insights – hot topics of research today – come in.

In 1981, Alec Jeffreys (born in 1950; he later became famous as the person who developed the technique of DNA 'finger-printing' used in forensic science) startled his colleagues with a discovery he announced at a meeting held at King's College, in Cambridge. By then, it had become clear that viruses could act as unwitting carriers of genetic information. When a phage invades a bacterium it uses the mechanisms of the bacterial cell to make copies of itself. In this process, it is quite easy for a bit of bacterial DNA to get copied into the 'new' viruses by mistake. When these viruses invade other cells, any cells that survive the infection might be left with an extra bit of DNA. In the overwhelming majority of cases, this would be ignored. And in any case, shifting a bit of DNA from one bacterium to another member of the same species doesn't imply any spectacular changes. But suppose the copied DNA could be moved from one species to another, and then be put to work by the cells of the second species? That is exactly what Jeffreys suggested to his colleagues in Cambridge.

He drew their attention to a protein called leghaemoglobin, which is used by plants known as legumes when they take nitrogen from the air and 'fix' it in the chemicals of their bodies. This is a key process for life on Earth because it produces ammonia (NH_3), which is essential for the manu-facture of amino acids, proteins and nucleic acids. The nitrogen compounds we need come only from our food; we cannot fix it ourselves. The actual process of nitrogen fixation takes place in bacteria, but in legumes these bacteria live in a symbiotic relationship with the cells of the plant. Jeffreys pointed out that the gene which codes for leghaemoglobin is, as the name suggests, very similar to the gene that codes for haemoglobin, a protein that is involved in the transport of oxygen in the blood of animals. He suggested that, long ago, the ancestral form of the animal gene had been carried as a passenger by a virus into the ancestral form of the plant, a process known as horizontal gene transfer, and that it had been adapted to its new role there by natural selection. The idea of horizontal gene transfer harks back to the work of Griffith, involving the sharing of genetic information between pneumococci, and is now well established as an important mechanism in the evolutionary behaviour of simple organisms. It is known to be a major factor in the spread of antibiotic resistance among bacteria, and in the evolution of bacteria that 'feed' off insecticides, breaking the insecticides down and reducing their effectiveness. But it would not be possible to transfer working genes in this way from, say, an oak tree or an elephant to a human being (or the other way around), because our bodies are too complex. Intriguing though these discoveries are, they are of only secondary interest to us as human beings. But the second new insight into the nature of evolution definitely concerns us as individuals, as well as bringing the story of the understanding of evolution up to date.

One of the clearest ways to appreciate that there is more to life than the simple inheritance of genes is to look at human identical twins. Such individuals are identical in appearance because they have developed from a single fertilised egg that split in two. So they have exactly the same genetic inheritance. But what is much more interesting is the fact that identical twins are not really identical. We are not talking about the cases where twins have been separated at an early age and brought up differently, which enables biologists to study differences between genetic influences and the influence of the environment they are brought up in – 'nature' versus 'nurture'. Even twins that have stayed together since birth and have experienced the same outside influences grow up with differences. In some cases, this shows up as a different susceptibility to *inherited* diseases – for example, even if one twin develops Type I diabetes it is unlikely that the sibling will. Since both individuals have the same genes, something must be going on inside a cell that affects the way genes work.

At one level, this is no surprise. After all, since every cell in your body contains the same genetic information, there must be something going on which makes a liver cell, say, behave like a liver cell, while the cells of your skin behave like skin cells. People do not suddenly start growing livers all over their skin, although every skin cell contains the genetic information required to grow a liver. If part of the mechanism that makes liver cells work like liver cells goes awry in one individual, he or she might develop diabetes, even though their genes are unaltered. It isn't so different from the way blobs of 'wrong' colour appear in maize plants. But what may come as a surprise is just how much of the DNA in your cells seems to be involved in controlling the behaviour of the genes.

The two strands of DNA in a double helix are held together by pairs of bases, like the teeth in a zip. Biochemical techniques have now become sophisticated enough to identify all of the DNA in the human genome. Each cell in your body contains about 6,000,000,000 base pairs linked along strands of DNA. But out of all that DNA, only some 120,000,000 base pairs carry the code for making proteins. This is only about two per cent of the total. Roughly 98 per cent of the DNA in your cells is not involved in coding for proteins, and is sometimes referred to as 'non-coding' DNA as a result. For a time, following its identification, it was assumed that this DNA did nothing at all, and it was derogatively dismissed as 'junk DNA'. But a little thought shows how unlikely this is. Even within the cell, there is competition for resources, and evolution is at work. Cells that wasted most of their resources on useless DNA would be unlikely to win out in the struggle for survival compared with more efficient cells.

The fact that this extra DNA is important for the functioning of living organisms is highlighted by comparing the amount found in simple organisms with the amount found in more complex organisms. The amount of DNA that *does* code for proteins is, of course, greater in creatures like ourselves or mice than in, say, bacteria or yeast. But the *relative* proportion of 'non-coding' DNA in more complex organisms is greater still. In a bacterium, the proportion of DNA that does not code for protein is about 10 per cent. In a fruit fly, it is 82 per cent. But in our own cells, as we have mentioned, it is 98 per cent. The more complex an organism is, the greater the proportion of 'useless' DNA it carries in its cells.

Of course, this DNA is not really useless. The 'non-coding' DNA must be doing something – and clearly something important – even if it doesn't code for protein. Apart from anything else, it can make strands of RNA that do not translate the instructions into protein, but influence the workings of the cell. Judging from the evidence of the amount of DNA involved, operating a cell is much more complicated than operating a human body. But to get a grip on how this works we need to have a more detailed picture of what is going on in the nucleus of a cell, where the DNA is concentrated.

The nucleus is only about 10 microns (one-hundredth of a millimetre) across. Yet each cell in your body contains a total length of DNA that would stretch for about 1.8 metres if all the pieces were laid end to end. This is packed into 46 tiny cylinders (23 pairs of chromosomes), which would themselves have a total length of 0.2 millimetres if laid end to end. In round numbers, the DNA is packed into just one ten-thousandth of its 'natural' length.

It happens like this. There is a family of proteins, called histones, which provide the scaffolding on which DNA is tightly wound and packed into a small space. A cluster of eight histones fit together to make a shape like a bead, and a strand of DNA makes two loops around the bead, like a rope being

On the Origin of Evolution

wrapped around a basketball. Another histone sits over the strands, clamping the loops in place. On either side of the bead, a short stretch of 'spacer' DNA provides a link to the next bead (properly called a nucleosome), and because this link is flexible a whole string of nucleosomes can be coiled up to make a compact structure, which can then be 'supercoiled' into an even more compact form. This is a masterpiece of packaging. But it means that when the cell needs to use a particular piece of genetic information the relevant stretch of DNA has to be opened up just enough for the information to be copied onto messenger RNA, then packed away neatly back where it belongs. It turns out that histones are more than just scaffolding, and that they play a part in how this unpacking, reading and repacking of genes goes on. More than fifty different kinds of histone activity have so far been identified, some of which make it easier for genes to be read, some of which make it harder, and others which operate in more subtle ways. This is very much ongoing research, but it is sufficient for our purposes to know that histones are involved in activating or de-activating genes.

Another cell mechanism also controls the activity of genes. This is called methylation, because it involves chemical units known as methyl groups. These are small units of carbon, oxygen and hydrogen atoms, containing the methyl 'radical' CH_3, which can attach to DNA strands at specific places, where there are cytosine and guanine bases next to each other. Methylation typically acts as a 'silencer' for a gene, so in many cases a gene can be turned on by demethylation.*

In a neat link with the past, methylation has explained a phenomenon that puzzled Linnaeus. In the 1740s, he was shocked to see a variety of plant that looked like common

---

* Methylation has been likened to an 'on/off switch', while histones act more like an adjustable 'volume control'.

toadflax but had very different flowers. This was particularly disturbing to him because his classification system for plants was based on the appearance of the flowers; he wrote that it was 'no less remarkable than if a cow were to give birth to a calf with a wolf's head'. In the 1990s, the plant biologist Enrico Coen discovered that in these 'monstrous' plants a particular gene involved in determining the structure of the flower is smothered in methyl groups, so that it is inactive. This property is passed on through the seeds to subsequent generations.

Methylation can also affect strands of RNA, and there is a slightly more mysterious pattern of activity involving strands of RNA that float about inside the cell and modify histones or affect the activity of genes. But although the details of all these processes are far from being understood, the message to take away is that the genome is not set in one pattern of activity, and that even though the 'book of life' stays the same, which passages from the book are read out and acted upon depends on the circumstances in which the cell finds itself – the environment. This whole process of selecting which passages to act upon is known as epigenetics;* although there is no universally accepted precise definition of what the term means, that doesn't matter for our purposes.

An experiment involving mice highlights how this process can happen. There is a mouse strain in which an interesting pattern of hair colour is controlled by a single gene, called the *agouti* gene, or *a* for short. In normal agouti mice, the hair is black at the base, yellow in the middle and black at the end, because the *a* gene is only switched on during the middle of hair growth. But there is a mutant strain in which the offspring of a single pair of parents, in the same litter,

---

* Literally, 'above genetics', implying something more than genes themselves.

may have different hair colours – some all yellow, some all black, and some with intermediate shades. And the ratio of different colours among the offspring changes when the pregnant mice are fed a diet rich in foods that supply methyl groups. The nutrition of the mothers directly affects the colour of the babies' fur, by de-activating (or partially de-activating) the *agouti* gene. Such experiments cannot be carried out on human beings in scientific laboratories. But a couple of historical examples have shown that not only can the diet of pregnant human mothers affect the activity of the genes of their children, but, in a revelation that startled biologists, these influences could extend into subsequent generations – they are inherited. Even more startlingly, a similar effect occurs when it is the fathers who are subjected to extreme diets.

A grim example of epigenetics at work was provided by events near the end of World War II, in the European winter of 1944. In retaliation against premature celebrations and a railway strike by the Dutch, who mistakenly believed that they were about to be liberated by the advancing Allied armies, the Nazi German authorities deliberately withheld food supplies from the occupied territory. The situation was exacerbated by an unusually cold winter. Four-and-a-half million people were restricted to an allowance of just 580 calories of food per day, and more than 22,000 of them died of starvation in what became known as the Hunger Winter. Because the Netherlands was a country with a well-developed health care system and good medical records, this unplanned 'experiment' provided a mass of data for researchers studying the effect of famine on children born in its aftermath.

The first thing they noticed was that if the mother had plenty of food during the months immediately after the baby was conceived, but went hungry in the later stages of pregnancy, the baby was likely to be small. But if she was starving during the first three months of pregnancy, then had plenty of food, the baby would be the usual kind of weight. These babies had in effect caught up with the 'normal' pattern by growing faster when food was plentiful. This pattern continued as the children grew up. The babies that were born small stayed small, but the babies that had a late growth spurt while still in the womb had unusually high obesity rates in later life, as if their bodies were still trying to compensate for the malnutrition early in foetal development. And the heavier adults also suffered from various ailments that were less common among the underweight individuals, including schizophrenia. Clearly, something that happened during the early stages of foetal development had affected not the genetic blueprint itself, but the way that blueprint was being interpreted. This is not too surprising in itself, but what was surprising was that this effect was passed on further – the children of the underweight children (the grandchildren of the mothers who experienced the Hunger Winter) were also underweight, even though they and their mothers were well fed. When biologists worked out what was going on, they replicated this with worms. They confirmed that the effect could be passed on to subsequent generations – *heritable* epigenetics.

This particular effect was linked not with methylation but with the activity of small pieces of RNA that also influence the way genes are expressed. Some of these molecules, now known as 'starvation-response small RNAs', live up to their name by affecting the way the cells they inhabit process nutrition when food is in short supply. And once this response has been switched on, it stays switched on through at least three generations of worms, even when the offspring are not starved. Then came another bombshell. It is easy to understand that the nutrition of a human mother affects the development of a baby in her womb, even if it is surprising that influences of this kind may extend into subsequent generations. But nobody expected to find that the nutrition of the fathers influences the babies in this way. Yet that is exactly what is now known to be the case. The evidence comes from studies of the records of another famine, or series of famines, from an isolated community in northern Sweden at the end of the nineteenth century and early in the twentieth century. The history of the community shows a series of good years and bad years for the harvests, with families who had suffered during the famines naturally tending to feast during the good years.

This time, the medical records cover several generations of the descendants of the survivors of those famines. What emerges is a link between the fate of those descendants and the availability of food for their male ancestors during a pre-adolescent stage known as the slow growth period, or SGP, which is a usual pattern of human development in the years immediately preceding puberty. When a boy was badly nourished during the SGP, his *grandsons* tended to have a lower risk of dying from stroke, high blood pressure or heart disease. But if a boy had an abundance of food, especially meat and dairy produce, when there was a glut during his SGP, his grandsons had a greater risk of obesity and diseases such as diabetes; their life expectancy was about six years less than that of grandsons of boys who had only had limited food available in the run-up to puberty. Although this discovery was intriguing, it relied on old records and small numbers of people. So, naturally, researchers turned to laboratory studies to check out the implications.

In one study, male albino rats from a standard laboratory strain were given a high-fat diet, then allowed to mate with females that had been given a standard diet. The offspring had normal weight, but symptoms linked with diabetes. In another experiment, male mice were given a diet low in protein, but with extra sugar to make up the calories. Then they were mated with females that had been given a normal diet. This time, the researchers studied directly the activity of genes in the livers of the offspring, and again found changes of the kind linked with diabetes. All of this shows that environmental influences on males can produce epigenetic changes that are passed on to at least the next generation, and maybe further. The changes are *not* being produced by a change in the environment of a foetus while it is in the womb. In yet another study, female mice kept on a low-calorie diet while pregnant produced babies that were underweight and prone to diabetes. And even though these offspring were fed normally, their own babies were born small and had a higher risk of diabetes.

There are other influences that work in a similar fashion, but diet is a particularly interesting example that we have chosen to highlight because it is relevant to a major health problem today. Many countries are experiencing what is called an epidemic of obesity. It is a natural assumption that the children of overweight parents become overweight themselves because the parents give them too much food. But maybe they become overweight and suffer from things like diabetes at least partly because their fathers had too much to eat when they were themselves children. There may actually be just a grain of truth in the old excuse 'I can't help being fat, it's in my genes'. Not that this means overweight people can't try to rectify the situation by suitable diet and exercise, but it might help society get to grips with the problem. This is just one example of the practical benefits of understanding epigenetics, but to look further into such matters would go beyond the scope of the present book.

From our point of view, the message to take away is that the scientific understanding of evolution is itself still evolving as we enter the third decade of the twenty-first century.

Darwin and Wallace were right about the role of natural selection, but the flexibility associated with epigenetic control of the way proteins are expressed gives complex organisms a kind of leeway against disaster. When the environment changes, there is scope for an organism to change even without waiting for a useful mutation to occur. And if that leeway allows a species to survive for long enough, then there may be time for a beneficial mutation that would enable a new variety to thrive, not just struggle along, and possibly even to replace the ancestral form. There is still much to learn about all this. In her Nobel Lecture, McClintock referred to:

> . . . *unusual responses of a genome to various shocks it might receive, either produced by accidents occurring within the cell itself, or imposed from without, such as virus infections, species crosses, poisons of various sorts, or even altered surroundings such as those imposed by tissue culture.*

But as she also emphasised:

> *We know nothing, however, about how the cell senses danger and instigates responses to it that often are truly remarkable.*

The story of evolution may, it seems, be just beginning.

ACKNOWLEDGEMENTS

We thank the Alfred C. Munger Foundation for financial support, and the University of Sussex for providing us with a base from which to work.

ENDNOTES

1 Patricia Fara, *Science: A Four Thousand Year History*, Oxford UP, 2009, page 235.
2 Aristotle, *Physics*, quoted by Osborn.
3 See Conway Zirkle, *Natural Selection Before the 'Origin of the Species'*. Proceedings of the American Philosophical Association, volume 84, page 71, 1941.
4 Quoted by Osborn.
5 Quoted by Lisa Jardine, *The Curious Life of Robert Hooke*, HarperCollins, London, 2002.
6 See Drake.
7 See Raven.
8 Quoted by Blunt.
9 See Frängsmyr.
10 http://cogweb.ucla.edu/EarlyModern/Maupertuis_1745.html

11 See Wilson.
12 See Knight.
13 *Les Cabales*, first published 1772, facsimile available from Kessinger, 2010.
14 See Roger.
15 'Biographical account of the late James Hutton, M.D.', in *Works*.
16 See Winchester.
17 See Bowlby.
18 See Bowlby.
19 Quoted in *Lord Kelvin and the Age of the Earth*, J. Burchfield, Macmillan, London, 1975. Thomson became Lord Kelvin in 1892.
20 See *The Dating Game*, Cherry Lewis, Cambridge UP, 2000.
21 *The Collected Letters of Samuel Taylor Coleridge*, Clarendon Press, Oxford.
22 Translation from Jordanova.
23 See Rudwick.
24 Carl Gustav Carus, *The King of Saxony's Journey Through England and Scotland in the Year 1844*, Chapman & Hall, London, 1846.
25 See *The Life and Letters of the Rev. Adam Sedgwick*, Cambridge UP, 1890.
26 See Browne, *Charles Darwin: The Power of Place*.
27 Quoted by van Wyhe.
28 *A Narrative of Travels on the Amazon and Rio Negro*.
29 *Darwin on Man*, Wildwood House, London, 1974.
30 See Francis Darwin, *The Life and Letters of Charles Darwin*, John Murray, London, 1887; quote is from Volume II, page 197. Available at http://darwin-online.org.uk/content/
31 See Gruber.
32 Longman, London, 1986.

33 See Wallace, *My Life*.
34 See *Schwann and Schleiden Researches*, translated by H. Smith, Sydenham Society, 1847.
35 See Iltis.
36 See Judson.
37 *The Nature of the Chemical Bond*, Cornell UP, 1940.
38 Quoted by Judson.
39 http://www.the-scientist.com/?articles.view/articleNo/22403/title/Martha-Chase-dies/
40 See Sayre.
41 See Olby.
42 *Genetical Implications of the Structure of Deoxyribonucleic Acid*.

SOURCES AND FURTHER READING

There are two essential online resources for anyone interested in the origins of evolution:

http://darwin-online.org.uk/

and

http://wallace-online.org/

Both include biographical information and a mass of publications, correspondence and notes by the subjects. Many of the quotes we have used, especially in Chapters 5 and 6, are taken from here.

260 *On the Origin of Evolution*

Elizabeth Agassiz, *Louis Agassiz: His Life and Correspondence*, Houghton Mifflin, Boston, 1886 (in 2 volumes).

Claude Albritton, *The Abyss of Time*, Freeman, Cooper & Co., San Francisco, 1980.

Antoine-Joseph Dézallier d'Argenville, *L'Histoire Naturelle*, originally published 1757, available from Forgotten Books, 2018.

Svante Arrhenius, *Worlds in the Making*, Harper, New York, 1908.

John Baker, *Abraham Trembley of Geneva*, Edward Arnold, London, 1952.

Nora Barlow, editor, *Charles Darwin's Diary of the Voyage of H.M.S. 'Beagle'*, Cambridge UP, 1933.

Nora Barlow, editor, *Charles Darwin and the Voyage of the 'Beagle'*, Philosophical Library, New York, 1946.

Nora Barlow, editor, *The Autobiography of Charles Darwin*, complete edition, Norton, New York, 1958.

Barrett, P. H., Gautrey, P. J., Herbert, S., Kohn D. & Smith, S. editors. *Charles Darwin's Notebooks*, British Museum, London, 1987.

Henry Walter Bates, *The Naturalist on the River Amazons*, John Murray, London, 1892.

William Bateson, *Mendel's Principles of Heredity*, Cambridge UP, 1909 (includes a reprint of Mendel's classic paper).

David Beeson, *Maupertuis*, Oxford UP, 1992.

Jim Bennett, Michael Cooper, Michael Hunter and Lisa Jardine, *London's Leonardo*, Oxford UP, 2003.

Wilfrid Blunt, *Linnaeus*, Frances Lincoln, London, 2004.

Russell Bonduriansky & Troy Day, *Extended Heredity*, Princeton UP, 2018.

John Bowlby, *Charles Darwin*, Hutchinson, London, 1990.

Peter Bowler, *The Mendelian Revolution*, Athlone, London, 1989.

John Langdon Brooks, *Just Before the Origin*, Columbia UP, New York, 1984.

Janet Browne, *Charles Darwin: Voyaging*, Cape, London, 1995.

Janet Browne, *Charles Darwin: The Power of Place*, Cape, London, 2002.

Georges Buffon, *Natural History*, Strahan and Cadell, London, 1785, translated by William Smellie, available online at https://archive.org/details/naturalhistoryge02buffuoft.

Georges Buffon, *The Epochs of Nature*, translated and edited by Jan Zalasiewicz, Anne-Sophie Milon and Mateusz Zalasiewicz, University of Chicago Press, 2018.

Frederick Burkhardt and colleagues, editors, *The Correspondence of Charles Darwin*, Cambridge UP, 1985 onwards.

Thomas Burnet, *The Sacred History of the Earth*, originally published in Latin in two volumes, 1681 and 1689; available from Forgotten Books, 2018, or as download from https://orange36.com/wp-content/uploads/2013/01/The-Sacred-Theory-of-the-Earth-Books-123-and-4-from-1691-347-pgs.pdf.

James Burnett (Lord Monboddo), *Of the Origin and Progress of Language*, Balfour & Cadell, Edinburgh, six volumes published 1773 to 1792.

Samuel Butler, *Evolution, Old and New*, Hardwicke & Bogue, London, 1879.

Nessa Carey, *The Epigenetics Revolution*, Icon, London, 2011.

Robert Chambers, *Vestiges of the Natural History of Creation*, Churchill, London, 1844. Available as a British Library Historical Print Edition, 2011.

Robert Chambers, *Explanations: A Sequel to the 'Vestiges of the Natural History of Creation'*, Churchill, London, 1845.

Teilhard de Chardin, *The Phenomenon of Man*, English translation, Collins, London, 1959.

Ronald Clark, *JBS: The Life and Work of J. B. S. Haldane*, Oxford UP, 1984.

E. L. Cloyd, *James Burnett, Lord Monboddo*, Oxford UP, 1972.

William Coleman, *Georges Cuvier, Zoologist*, Harvard UP, 1964.

Nathaniel Comfort, *The tangled field: Barbara McClintock's search for the patterns of genetic control*, Harvard UP, 2001.

Georges Cuvier, *Lessons in Comparative Anatomy*, available as 'Cuvier's History of the Natural Sciences: twenty-four lessons from Antiquity to the Renaissance', translated by Abby S. Simpson, edited by Theodore Pietsch, Publications scientifiques du Muséum national d'Histoire naturelle, Paris, 2012.

Georges Cuvier, *Essay on the Theory of the Earth*, Blackwood, Edinburgh, 1813.

Cyril Darlington, *Darwin's Place in History*, Blackwell, Oxford, 1959.

Charles Darwin, *Journal of Researches*, Hafner, New York, 1952 (reprint of 1839 edition). Aka *Voyage of the Beagle*; see also Nora Barlow.

Charles Darwin, *On the Origin of Species*, John Murray, London, 1859. The first edition is the best and most reliable. This version is good: *The Annotated Origin: A Facsimile of the First Edition of On the Origin of Species*, Harvard UP, 2011. The 'Historical Sketch' which he included in later editions can be found here: http://oll.libertyfund.org/pages/darwin-s-historical-sketch-on-the-origin-of-species.

Charles Darwin, *Variation of Animals and Plants under Domestication*, John Murray, London, 1868 (in 2 volumes).

Charles Darwin, *The Descent of Man*, John Murray, London, 1871, available as a Penguin Classic, London, 2004.

Erasmus Darwin, *The Botanic Garden*, Johnson, London, 1791.

Erasmus Darwin, *Zoonomia*, Johnson, London, in 2 volumes, 1794/1796.

Erasmus Darwin, *The Temple of Nature*, Johnson, London, 1803.

Francis Darwin, editor, *Life and Letters of Charles Darwin*, John Murray, London, 1887 (in 3 volumes).

Francis Darwin, editor, *Foundations of the Origin of Species*, Cambridge UP, 1909.

Francis Darwin and A. C. Seward, editors, *More Letters of Charles Darwin*, John Murray, London, 1903 (in 2 volumes).

Richard Dawkins, *The Selfish Gene*, Oxford UP, 1976.

Richard Dawkins, *The Extended Phenotype*, Freeman, Oxford, 1982.

Dennis Dean, *James Hutton and the History of Geology*, Cornell UP, 1992.

Daniel Dennett, *Darwin's Dangerous Idea*, Simon & Schuster, New York, 1995.

Adrian Desmond, *Huxley*, Penguin, London, 1997.

Theodosius Dobzhansky, *Genetics and the Origin of Species*, Columbia UP, 1937.

Ellen Tan Drake, *Restless Genius: Robert Hooke and His Earthly Thoughts*, Oxford UP, 1996.

Lee Alan Dugatkin, *Mr. Jefferson and the Giant Moose: Natural History in Early America*, University of Chicago Press, 2009.

Loren Eiseley, *Darwin's Century*, Doubleday, New York, 1958.

Niles Eldredge, *Time Frames*, Heinemann, London, 1986.

Georgina Ferry, *Dorothy Hodgkin*, Bloomsbury, London, 2014.

Martin Fichman, *An Elusive Victorian: The Evolution of Alfred Russel Wallace*, Chicago UP, 2004.

R. A. Fisher, *The Genetical Theory of Natural Selection*, Clarendon Press, Oxford, 1930.

Tore Frängsmyr, editor, *Linnaeus*, University of California Press, Berkeley, 1983.

Francis Galton, *Natural Inheritance*, Macmillan, London, 1889.

Etienne Geoffroy Saint-Hilaire, *Philosophie anatomique*, published in 2 volumes 1818–1822, reprint from Ulan Press, 2012, via Amazon.

Charles Gillispie, *Genesis and Geology*, Harvard UP, 1951.

Stephen Jay Gould, *Wonderful Life*, Hutchinson Radius, London, 1989.

Stephen Jay Gould, *The Structure of Evolutionary Theory*, Belknap, Harvard, 2002.

John Gribbin, *In Search of Schrödinger's Cat*, Bantam, London, 1984.

John Gribbin, *Science: A History*, Allen Lane, London, 2002.

John Gribbin, *The Cosmic Origins of Life*, Endeavour Press, 2019.

John Gribbin & Jeremy Cherfas, *The First Chimpanzee*, Penguin, London, 2001.

Mary Gribbin & John Gribbin, *Flower Hunters*, Oxford UP, 2008.

Howard Gruber, *Darwin on Man*, Wildwood House, London, 1974.

Robert Gunther, *Further Correspondence of John Ray*, Ray Society, London, 1928.

Ernst Haeckel, *The History of Creation*, Appleton, New York, 1880 (in 2 volumes); originally published in German, 1868.

Knut Hagberg, *Carl Linnaeus*, Dutton, New York, 1953.

J. B. S. Haldane, *The Causes of Evolution*, originally published 1932, available in a nice edition from Princeton UP, 1990.

Robin Henig, *A Monk and Two Peas: The Story of Gregor*

Mendel and the Discovery of Genetics, Phoenix, London, 2001.

Sandra Herbert, *Charles Darwin, Geologist*, Cornell UP, Ithaca, 2005.

Lewis Hicks, *A Critique of Design Arguments*, Scribner's, New York, 1883; available at https://archive.org/details/critiqueofdesign00hick.

Jonathan Hodge, *Before and After Darwin*, Routledge, London, 2008.

Jonathan Hodge and Gregory Radick, editors, *The Cambridge Companion to Darwin*, Cambridge UP, 2009.

Robert Hooke, *Micrographia*, Royal Society, London, 1665; facsimile edition Dover, New York, 1961.

Robert Hooke, *Lectures and Discourses on Earthquakes*, reprinted from the Posthumous Works (edited by Richard Waller, 1705) and published by Arno Press, New York, 1978.

Alexander von Humboldt, *Personal Narrative of Travels*, Longmans, Hurst, Orme, Rees and Brown, London, 1814. Available as a Penguin Classic abridged edition.

Michael Hunter and Simon Schaffer, editors, *Robert Hooke: New Studies*, Boydell Press, Woodbridge, 1989.

James Hutton, *Theory of the Earth*, Creech, Edinburgh, 1795 (in 3 volumes); originally published 1788 in the Transactions of the Royal Society of Edinburgh.

Julian Huxley, *Evolution: The Modern Synthesis*, Allen & Unwin, London, 1942.

Leonard Huxley, editor, *Life and Letters of Thomas Henry Huxley*, Macmillan, London, 1913 (in 3 volumes).

Leonard Huxley, *Life and Letters of Sir Joseph Dalton Hooker*, John Murray, London, 1918 (in 2 volumes).

Thomas Henry Huxley, *Evidence as to Man's Place in Nature*, Williams & Norgate, London, 1863; available at www.gutenberg.org/files/2931/2931-h/2931-h.htm.

Thomas Henry Huxley, Collected Essays, nine volumes, available at https://archive.org/details/collectedessays00huxl.

Hugo Iltis, *Life of Mendel*, Norton, New York, 1932.

Stephen Inwood, *The Man Who Knew Too Much*, Macmillan, London, 2002.

Roland Jackson, *The Ascent of John Tyndall*, Oxford UP, 2018.

Lisa Jardine, *The Curious Life of Robert Hooke*, HarperCollins, London, 2003.

Ludmilla Jordanova, *Lamarck*, Oxford UP, 1984.

Horace Freeland Judson, *The Eighth Day of Creation*, Cape, London, 1979.

Immanuel Kant, *Universal Natural History and Theory of the Heavens*, originally published in German, 1755, available from Scottish Academic Press, Edinburgh, 1981.

Evelyn Fox Keller, *A Feeling for the Organism*, Freeman, San Francisco, 1983.

Paul Kent and Allan Chapman, editors, *Robert Hooke and the English Renaissance*, Gracewing, Leominster, 2005.

Geoffrey Keynes, *A Bibliography of Dr Robert Hooke*, Clarendon Press, Oxford, 1966.

Desmond King-Hele, editor, *The Essential Erasmus Darwin*, McGibbon & Kee, London, 1968.

Desmond King-Hele, *Erasmus Darwin*, De la Mare, London, 1999.

William Knight, *Lord Monboddo*, John Murray, London, 1900.

Ernst Krause, *Erasmus Darwin*, Appleton, New York, 1880.

Jean-Baptiste Lamarck, *Zoological Philosophy*, originally published in French, 1809, now available in English from Forgotten Books, London, 2016 (see https://www.forgottenbooks.com/en).

Nick Lane, *Life Ascending*, Profile, London, 2010.

Nick Lane, *The Vital Question*, Profile, London, 2015.

Edwin Lankester, *The Correspondence of John Ray*, Ray Society, London, 1848. Available at https://archive.org/stream/correspondenceof48rayj#page/n13/mode/2up.

Pierre-Simon Laplace, *The System of the World*, English translation originally published 1809, available at https://archive.org/details/systemworld01laplgoog.

William Lawrence, *Lectures on physiology, zoology and the natural history of man*, Callow, London, 1819.

William Leonard, *The Fragments of Empedocles*, Open Court, Chicago, 1908.

Cherry Lewis, *The Dating Game*, Cambridge UP, 2000.

Richard Lewontin, *The Genetic Basis of Evolutionary Change*, Columbia UP, 1974.

Lucretius, *The Nature of Things*, Penguin, London, 2007.

Charles Lyell, *Principles of Geology*; originally in three volumes from John Murray, London, but available now as a single volume from Penguin, London, 1997.

Charles Lyell, *Geological Evidences of the Antiquity of Man*, originally published by John Murray, London, 1863, extensively revised 1873.

Katherine Lyell, editor, *Life, Letters and Journals of Sir Charles Lyell*, John Murray, London, 1881 (in 2 volumes).

Cherrie Lyons, *Thomas Henry Huxley*, Prometheus, New York, 1999.

Benoît de Maillet, *Telliamed*, originally published 1748, trans. Albert Carozzi, University of Illinois Press, Urbana, 1968.

Thomas Malthus, *An Essay on the Principles of Population*, short version published 1798, fuller version 1803. Now available as a Penguin Classic, London, 2015.

James Marchant, *Alfred Russel Wallace*, Harper, New York, 1916.

Lynn Margulis, *Symbiosis in Cell Evolution*, Freeman, New York, second edition 1993.

Patrick Matthew, *On Naval Timber and Arboriculture*, Adam Black, Edinburgh, 1831.

Pierre-Louis Moreau de Maupertuis, *Ouvres*, originally published in 4 volumes in 1768, reprinted by Olms, Hildesheim, 1968; available at https://archive.org/details/uvresdemaupertui01maup.

Ernst Mayr, *Evolution and the Diversity of Life*, Harvard UP, 1976.

Ernst Mayr, *The Growth of Biological Thought*, Harvard UP, 1981.

Ernst Mayr, *What Evolution Is*, Basic Books, New York, 2001.

H. L. McKinney, *Wallace and Natural Selection*, Yale UP, 1972.

Gregor Mendel, *Experiments on Plant Hybridization*, Harvard UP, 1965.

Milton Millhauser, *Just Before Darwin: Robert Chambers and the Vestiges*, Wesleyan UP, Middletown, Conn., 1959.

Mondoddo (Lord), see Burnett (James).

Thomas Hunt Morgan, *Evolution and Adaptation*, originally published 1903, available from Cornell UP, 2009.

Simon Conway Morris, *The Crucible of Creation*, Oxford UP, 1998.

Robert Olby, *The Path to the Double Helix*, Macmillan, London, 1974.

Alexander Ivanovich Oparin, *The Origin of Life*, Macmillan, London, 1938; reprinted by Dover, New York, 1953.

Vitezslav Orel, *Gregor Mendel*, Oxford UP, 1995.

Henry Fairfield Osborn, *From the Greeks to Darwin*, Macmillan, London, 1894 (second edition 1902).

Dorinda Outram, *Georges Cuvier*, Manchester UP, 1984.

Richard Owen, *On the Archetype and Homologies of the Vertebrate Skeleton*, van Voorst, London, 1848.

Richard Owen, *On the Nature of Limbs*, van Voorst, London, 1849; University of Chicago Press edition, 2008.

Alpheus Packard, *Lamarck, the founder of Evolution*, Longmans, New York, 1901.

William Paley, *Natural Theology*, originally published 1822, available as an Oxford UP Classic, 2006.

John Playfair, *Illustrations of the Huttonian Theory of the Earth*, originally published 1802, British Library Historical Print Edition, 2011.

John Playfair, *The Works of John Playfair*, Constable, Edinburgh, 1822.

Roy Porter, *The Making of Geology*, Cambridge UP, 1977.

Peter Raby, *Alfred Russel Wallace*, Pimlico, London, 2002.

Charles Raven, *John Ray*, Cambridge UP, 1942.

John Ray, *Miscellaneous Discourses Concerning the Dissolution and Changes of the Earth*, originally published by Samuel Smith, London, 1692, reprinted by Olms, Hildesheim, 1968.

Jacques Roger, *Buffon*, Cornell UP, Ithaca, 1997.

Matt Rossano, *Supernatural Selection: How Religion Evolved*, Oxford UP, 2010.

Martin Rudwick, *Georges Cuvier*, University of Chicago Press, 1997.

Anne Sayre, *Rosalind Franklin and DNA*, Norton, New York, 1975.

Erwin Schrödinger, *What is Life?*, first published 1944, available in a single volume with his book *Mind and Matter*, Cambridge UP, 1967.

George Scrope, *Considerations on Volcanoes*, Phillips, London, 1825.

James Secord, *Victorian Sensation: The Extraordinary Publication, Reception, and Secret Authorship of Vestiges of the Natural History of Creation*, University of Chicago Press, 2000.

George Gaylord Simpson, *The Major Features of Evolution*, Columbia UP, 1953.

Charles Smith, editor, *Alfred Russel Wallace: An Anthology of his Shorter Writings*, Oxford UP, 1991.

John Maynard Smith, *Evolution and the Theory of Games*, Cambridge UP, 1982.

Nicholas Steno, *Prodromus* (The Prodromus of Nicolaus Steno's Dissertation Concerning a Solid Body Enclosed by Process of Nature Within a Solid), originally published 1669, translation by John Winter, with notes, University of Michigan Press, 1916.

Jenny Uglow, *The Lunar Men: The Friends Who Made the Future 1730–1810*, Faber & Faber, London, 2002.

Leonardo da Vinci, *The Notebooks of Leonardo da Vinci* (translated by E. MacCurdy in two volumes), Cape, London, 1938.

Hugo de Vries, *Species and Varieties*, edited by Daniel MacDougal, Open Court, Chicago, 1905; available at https://archive.org/details/speciesvarieties00vrieuoft.

Hugo de Vries, *The Mutation Theory*, Open Court, Chicago, 1910.

Alfred Russel Wallace, *A Narrative of Travels on the Amazon and Rio Negro*, Reeve, London, 1853.

Alfred Russel Wallace, *The Geographical Distribution of Animals*, Harper & Brothers, New York, 1876; Kindle edition available at Amazon.

Alfred Russel Wallace, *Darwinism: An Exposition of the Theory of Natural Selection with Some of Its Applications*, Macmillan, London, 1889; available at https://archive.org/stream/darwinismexposit00walluoft#page/n5/mode/2up.

Alfred Russel Wallace, *My Life*, Chapman & Hall, London, 1905.

Alfred Russel Wallace, *Letters from the Malay Archipelago*, edited by John van Wyhe & Kees Rookmaaker, Oxford UP, 2015.

Richard Waller, editor and publisher, *The Posthumous Works of Robert Hooke*; originally published 1705, available at https://play.google.com/store/books/details/Robert_Hooke_The_Posthumous_Works_of_Robert_Hooke?id=6xVTA AAAcAAJ.

Jonathan Weiner, *The Beak of the Finch*, Knopf, New York, 1994.

August Weismann, *Essays upon Heredity*, Clarendon Press, Oxford, 1891–1892 (in 2 volumes).

August Weismann, *On Germline Selection*, originally published 1896, in English by Open Court, Chicago, 1902.

August Weismann, *The Evolution Theory*, Edward Arnold, London, 1904 (in 2 volumes); available at https://archive.org/details/evolutiontheory02weis_0.

William Wells, *Two Essays*, Constable, London, 1818.

Alfred North Whitehead, *Science and the Modern World*, originally published 1925, available from Cambridge UP, 2011.

George C. Williams, *Adaptation and Natural Selection*, Princeton UP, new edition, 1996.

Arthur Wilson, *Diderot*, Oxford UP, 1972.

Edward O. Wilson, *Sociobiology*, Harvard UP, 1975.

Simon Winchester, *The Map that Changed the World*, HarperCollins, New York, 2001.

Sewall Wright, *Evolution: Selected Papers*, University of Chicago Press, 1986.

John van Wyhe, *Dispelling the Darkness*, World Scientific, Singapore, 2013.

Carl Zimmer, *Evolution*, HarperCollins, London, 2001.

INDEX